Studies in Computational Intelligence

Volume 1291

The series "Studies in Computational Intelligence" (SCI) publishes new developments and advances in the various areas of computational intelligence—quickly and with a high quality. The intent is to cover the theory, applications, and design methods of computational intelligence, as embedded in the fields of engineering, computer science, physics and life sciences, as well as the methodologies behind them. The series contains monographs, lecture notes and edited volumes in computational intelligence spanning the areas of neural networks, connectionist systems, genetic algorithms, evolutionary computation, artificial intelligence, cellular automata, self-organizing systems, soft computing, fuzzy systems, and hybrid intelligent systems. Of particular value to both the contributors and the readership are the short publication timeframe and the world-wide distribution, which enable both wide and rapid dissemination of research output.

Indexed by SCOPUS, DBLP, WTI AG (Switzerland), zbMATH, SCImago.

All books published in the series are submitted for consideration in Web of Science.

Abdelmounim Bouziane · Karima Bouziane
Editors

AI, Management, and Culture: Shaping Organizations in the Digital Age

Navigating the Intersection of Technology, People, and Performance

Springer

Editors
Abdelmounim Bouziane
Chouaib Doukkali University
El Jadida, Morocco

Karima Bouziane
Faculty of Letters and Humanities
Chouaib Doukkali University
El Jadida, Morocco

ISSN 1860-949X ISSN 1860-9503 (electronic)
Studies in Computational Intelligence
ISBN 978-3-032-27881-4 ISBN 978-3-032-27882-1 (eBook)
https://doi.org/10.1007/978-3-032-27882-1

This Springer imprint is published by the registered company Springer Nature Switzerland AG
The registered company address is: Gewerbestrasse 11, 6330 Cham, Switzerland

This book is dedicated to scholars, practitioners, and students who seek to understand and responsibly shape the transformative role of artificial intelligence in contemporary organizations.
It is also dedicated to those who recognize that technological innovation must be accompanied by managerial insight, cultural awareness, and ethical responsibility in the digital age.

Preface

Artificial intelligence has become a defining force in the transformation of modern organizations. Its applications now extend across management, communication, professional services, education, and public administration, profoundly influencing how organizations operate and interact with their environments.

This edited volume adopts an interdisciplinary and culturally informed perspective on artificial intelligence. Rather than viewing AI solely as a technological advancement, the contributions examine its integration as a socio-technical process shaped by managerial practices, organizational cultures, and institutional contexts.

Covering themes such as AI-assisted translation, branding and communication, auditing and control, education, and the digitalization of public services, the book offers both conceptual frameworks and empirical insights. It emphasizes the importance of governance, ethics, and cultural alignment in ensuring that AI-driven transformation remains sustainable and socially responsible.

This volume is intended for researchers, doctoral students, practitioners, and policymakers interested in understanding how artificial intelligence is reshaping organizations in the digital age.

The rapid diffusion of artificial intelligence has significantly altered organizational practices across both private and public sectors. AI technologies are no longer confined to technical experimentation; they now play a central role in decision-making, professional judgment, service delivery, and organizational communication.

This edited volume, AI, Management, and Culture: Shaping Organizations in the Digital Age, originates from a shared conviction that the impact of AI cannot be fully understood through a purely technological lens. Organizational transformation is shaped by the interaction between technology, management, culture, and human agency.

The chapters assembled in this book explore AI applications in diverse organizational contexts, including translation and intercultural communication, branding and marketing, auditing, education, and the digital transformation of public services. Collectively, they address critical issues such as technology acceptance, trust, accountability, ethical considerations, and cultural adaptation.

By bringing together contributions from different disciplines and methodological traditions, this volume aims to bridge theory and practice. It provides readers with a comprehensive understanding of AI as a driver of organizational change while highlighting the need for human-centered and ethically grounded management approaches in the digital era.

El Jadida, Morocco

Abdelmounim Bouziane
Karima Bouziane

Acknowledgements

The editors would like to express their sincere appreciation to all contributing authors for their scholarly engagement, rigor, and cooperation throughout the editorial process. Their work constitutes the core strength of this volume.

The editors also thank the reviewers for their insightful comments and constructive feedback, which significantly enhanced the quality and coherence of the chapters.

Gratitude is extended to the Springer editorial and production teams for their guidance, professionalism, and support at every stage of the publication process.

Finally, the editors warmly acknowledge the encouragement and support of their colleagues, institutions, families, and friends, whose patience and understanding made the completion of this book possible.

Contents

Contributors

Sahib Eddine Abdelhak ENCGJ, Chouaib Doukkali University, El Jadida, Morocco

Arare Ahmed ENCGJ, Chouaib Doukkali University, El Jadida, Morocco

Hiba Argoubi Euromed Business School, Euromed University of Fes, Fes, Morocco

Mohamed Assoud ISGA Rabat, Rabat, Morocco;
Hassan First University of Settat, Settat, Morocco

Badr Bentalha National School of Business and Management, Sidi Mohamed Ben Abdellah University, Fes, Morocco

Rachid Boudri Euromed Business School, Euromed University of Fes, Fes, Morocco

Abdelmounim Bouziane Chouaib Doukkali University & Urban Agency of Casablanca, Casablanca, Morocco

Karima Bouziane ALCS Lab, Faculty of Letters and Humanities, Chouaib Doukkali University, El Jadida, Morocco

Abdelkrim Chirig Chouaib Doukkali University, El Jadida, Morocco

Walid Dalal Faculty of Legal, Economic, and Social Sciences, LARGESS, Chouaib Doukkali University, El Jadida, Morocco

Mouna Darouassi Politique Educatifs et Dynamiques Sociales (PEDS), Faculty of Education Sciences (FSE), Mohamed 5 University, Rabat, Morocco;
Faculty of Education Sciences (FSE) PhD candidate and teaching monitor (monitrice) PEDS Laboratory, Mouna Darouassi Mohammed V University, Rabat, Morocco

Oumelbanine El Gueddar Politique Educatifs et Dynamiques Sociales (PEDS), Faculty of Education Sciences (FSE), Mohamed 5 University, Rabat, Morocco;

Faculty of Education Sciences (FSE) PEDS Laboratory, Oumelbanine El Gueddar, Rabat, Morocco

Samah Fifani Chouaib Doukkali University, El Jadida, Morocco

Taouil Halima ENCGJ, Chouaib Doukkali University, El Jadida, Morocco

Khalil Hamidi Applied Language and Culture Studies Laboratory, Faculty of Letters and Humanities, Chouaib Doukkali University, El Jadida, Morocco, El Jadida, Morocco

Marouane Ikira Faculty of Legal, Economic, and Social Sciences, LARGESS, Chouaib Doukkali University, El Jadida, Morocco

Hafssa Laabdi Applied Language and Culture Studies Laboratory, Faculty of Letters and Humanities, Chouaib Doukkali University, El Jadida, Morocco, El Jadida, Morocco

Mouad Mbeker Applied Language and Culture Studies Laboratory, Faculty of Letters and Humanities, Chouaib Doukkali University, El Jadida, Morocco, El Jadida, Morocco

Dounia Rabhi Faculty of Legal, Economic, and Social Sciences, LARGESS, Chouaib Doukkali University, El Jadida, Morocco

Amira Sebbah Hassan First University of Settat, Settat, Morocco

Artificial Intelligence in Tourism Management: Cultural and Strategic Comparisons Between Morocco and Spain

Taouil Halima, Arare Ahmed, and Sahib Eddine Abdelhak

Abstract The rapid advancement of AI significantly changes how the tourism industry operates. Currently the majority of studies have focused on the complex technological nature of AI adoption and have not taken into account the significant cultural and managerial influences on understanding, implementing and experiencing AI. The conceptual model outlined in this study establishes an important link between culture, managerial orientation, AI adoption and tourism performance by comparing data collected from Morocco and Spain. Based on a cross-cultural framework, the literature on tourism management, as well as recent empirical studies related to AI adoption, the authors propose that the way in which AI is integrated into tourism is not culturally neutral, nor can AI be viewed simply as a technological tool for service delivery. Rather, AI and its integration will be impacted by the types of services and governance framework in a given culture and the capacity of the cultural institution to support the use of AI. The authors suggest that while cultures which emphasize long-term relational perspectives will prefer to use AI in a manner which augments the experience of human services through personal connection and support for workers; cultures that are more structured and performance oriented will likely support and promote the institutionalization of AI as a strategic management system for SMEs.

Keywords Cultural context · Digital transformation · E-tourism ecosystems · Institutional capacity managerial orientation · Institutional capacity

1 Introduction

Over the last ten years, a growth of artificial intelligence (AI) in the tourism industry has occurred due to improvements in technology. AI allows for more efficient Tourist related activities. AI has been used by the tourism sector to facilitate various tasks such as pricing; forecasting demand; managing visitors and interacting with

T. Halima (✉) · A. Ahmed · S. E. Abdelhak
ENCGJ, Chouaib Doukkali University, El Jadida, Morocco
e-mail: taouil.halima@ucd.ac.ma

A. Bouziane and K. Bouziane (eds.), *AI, Management, and Culture: Shaping Organizations in the Digital Age*, Studies in Computational Intelligence 1291,
https://doi.org/10.1007/978-3-032-27882-1_1

customers. Earlier studies focused on the financial savings associated with AI implementation, while newer studies have indicated the importance of organizational culture, leadership styles and ways of doing business (managerial logic) towards the successful adoption/implementation of AI within the tourism industry (Buhalis and Sinarta 2019), suggestion by several academics that the pathway towards AI adoption/implementation within service industries may be different from service sector industry to service sector industry, as the way of management and institutional structures will affect how each business adopts/implements AI (Dwivedi et al. 2021). The effect of these factors (workforce norms; governance structures; expectations of society towards automation) will become more elaborated/descriptive when comparisons take place on a global scale such as between countries that have different cultural focus or ways of managing businesses. Through this research, this gap has been filled by establishing a conceptual framework that places managerial orientation, AI adoption and tourism outcomes within a broader cultural framework (Viglia et al. 2018) and in addition to applying such an approach to both Morocco and Spain. The application of this framework to the tourism context of both countries illustrates the importance of developing a contextualized understanding of AI adoption in tourism. By comparing and contrasting these two cultural contexts, this research has the potential to stimulate future empirical research so for that these are the main questions:

- How does the proposed conceptual framework link culture, managerial orientation, AI adoption, and tourism outcomes in the study?
- How do relational, high-context cultures differ from structured, performance-oriented contexts in their adoption and use of AI in tourism?

2 Literature Background

2.1 AI Definition in Tourism

Tourism industry uses the term Artificial Intelligence (AI) to refer to tools that are based on computers and can learn and act without human involvement to help managers with their decision-making, automate processes, and improve the overall visitor experience through the analysis of data (Huang and Rust 2018). In addition to the traditional IT systems used in tourism, AI also uses machine learning, natural language processing (NLP), and data-driven algorithms to develop the ability for systems to operate autonomously or semi-autonomously in unpredictable situations (Tussyadiah 2020). Through AI, tourism managers are able to utilize a large amount of data (for example, booking history, review history, location history, user activity, etc.) to derive insights and to develop personalized solutions for guests (López-Naranjo et al. 2025). Consequently, AI can be viewed as both a new technology and a new management paradigm that is radically changing how the tourism industry is organized, provided and delivered (Ghesh et al. 2023). A recent trend in tourism management is the use of generative AI technologies to develop content for visitors and to build tools and applications for travel agencies and travel companies to assist with planning, booking and decision-making throughout the entire process of visiting a new place (Li et al. 2025).

AI in tourism involves service automation by utilizing intelligent systems. Chatbots, virtual assistants, facial recognition check-in systems, and service robots perform routine and repetitive service tasks with little or no human intervention. In addition, predictive analytics is a key application for AI in Tourism as it enables tourism stakeholders to predict tourist demand, forecast seasonal changes, profile customers' preferences and create personalized travel experiences using machine learning and deep learning algorithms (Sousa et al. 2024). The hospitality and tourism industries benefit from AI implementation in several ways, including providing managers with access to more accurate and timely information through data analytics; streamlining operations through automation; facilitating new ways of working with employees; and providing the necessary tools for creating strategic plans to successfully implement intelligent systems within their service delivery and business processes (Shin et al. 2025) (Fig. 1).

Through its depiction of AI as a Strategic connection between the tourism industry's ability to function more effectively and the requirement to be culturally sustainable, this figure shows how intelligent management systems can bring forth both efficiency in the Tourism Industry and outcomes realized by Heritage-focused entities. AI technology allows for enhanced operational capabilities and pricing methods, as well as expanded options for automating services that support the operational functions of a Destination (Sánchez-Martín et al. 2025).

2.2 *Linking AI, Tourism Management and Culture*

Adopting AI technology does not involve a culture-neutral basis, but rather the cultural surroundings influence the way management may trust, use, and accept AI to support managerial decisions. Research shows that culture supports or prevents these forms of acceptance by creating differences between cultural contexts in terms of how much faith and confidence is placed in algorithms. Cultures characterized as being high in uncertainty avoidance necessitate more significant levels of transparency and human intervention in the deployment of AI technology than do cultures defined as having low levels of uncertainty avoidance (Chi et al. 2023). Moreover, variations exist between countries regarding their acceptance of automation and AI; collectivist cultures tend to be more willing than individualist cultures to perceive AI as a complement to and extension of human behavior, while vice versa, individualist cultures tend to perceive AI as more of a competitor, with a greater focus on protecting the human element in the workplace (Yam et al. 2023). Finally, culture regularizes the patterns by which managers make decisions related to AI in terms of whether an organization will utilize AI strategically or operationally; the systematic

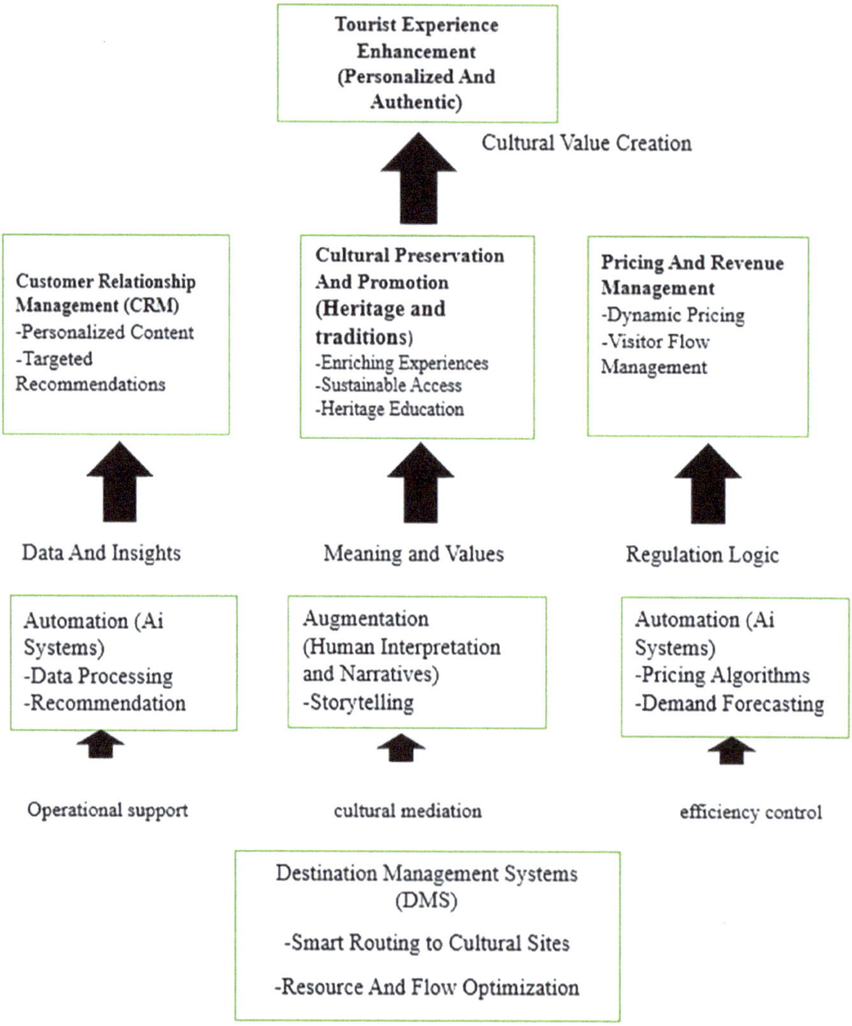

Fig. 1 AI framework for cultural preservation in tourism. *Source* Authors own elaboration based on Shin et al. (2025), Henriques and Pereira (2024), Buhalis and Leung (2018), Sousa et al. (2024), Huang and Rust (2018), Sharma (2024)

review of AI and strategic management literature identifies the ways cultural priorities dictate the importance of AI usages, as well as how the use of AI in organizations is achieved from a governance perspective (Keding 2021). Therefore, these results further substantiate how the manner in which AI is adopted is dependent upon a culture rather than being represented by a purely technological or culture-neutral approach (Fig. 2).

The figure displays how Generative AI (GenAI) connects disparate elements of tourism into one integrated system. The components connected by Gen AI

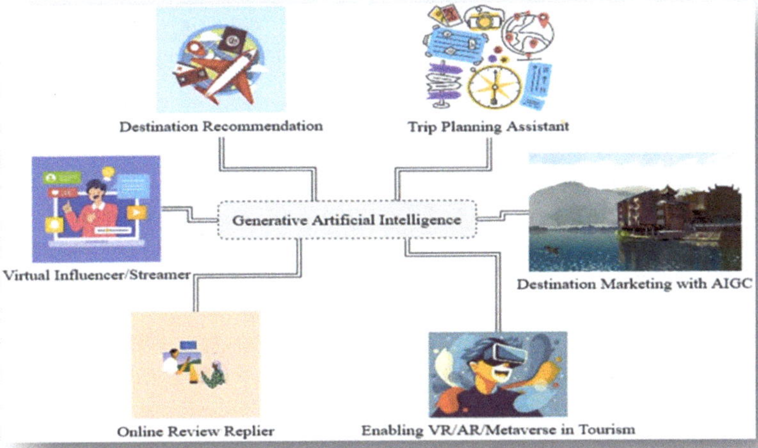

Fig. 2 Generative artificial intelligence as an integrated ecosystem for smart tourism management and experience design. *Source* Mengying et al. (2026)

include destination recommendation engines, trip-planning assistance applications, AI-generated marketing materials, virtual influencers, automated online review management systems and immersive VR/AR/metaverse technology. The integration of these disparate components creates opportunities for personalization, real-time destination management, scalable digital marketing, reputation management, and co-creating immersive experiences. The structure of the integrated GenAI components showcases GenAI as a strategic and systemic enabler of smart tourism by embedding technology-based innovations into the organizational capabilities, governance structures, and experience design to improve efficiency, competitiveness, and sustainability within tourism destinations (Buhalis and Sinarta 2023; Gretzel et al. 2023).

3 AI in the Moroccan Tourism Management

Digital technologies offer the possibilities of enhancing the Cultural-Tourism Connection. However, how much AI technology will actually impact Tourism in Morocco will depend on Cultural Values, as well as the organizational environment, in that these two areas influence the degree of acceptance and use of the technology. Research has shown that Culture will affect Technology Adoption by influencing Perceived Usefulness and Ease of Use, as well as Organizational and National Focus on Innovation, meaning that Culture moderates the success of Digital

Business Models in Morocco (Jan et al. 2024). In addition to the impact of AI on workers' capability & strategies, several studies have been conducted using various Versions of an Organization's IT/Technology Acceptance Framework; these studies show that Digital Culture & Competencies mediate the impact of AI upon workers' capabilities & success. There is substantial evidence that the shared Values & Competencies of an organization will significantly impact how an organization integrates AI into their Tourism Management Processes examine the importance of Cultural Behaviors and Factors relative to the "Usefulness" of AI applications in Tourism (Emerging Markets) during the process of their adoption. In addition to Technology, Community Values and Local Business Practices have an influence on the rate of diffusion of technology, in particular as it pertains to AI and Digital Platforms within the Tourism Industry. Consequently, managers who are operating in Morocco must develop cultural intelligence and build capacity within their organizations in order to bridge the Digital Divide and promote Greater AI Integration amongst their stakeholders (Abdullah Alhelal et al. 2025) (Fig. 3).

The Ministry of Tourism in Morocco has historically developed its policies based on what is required by state intervention and subsequently through Vision 2010 and Vision 2020s long-term plans to balance cultural heritage with economic growth (OECD, OECD tourism trends and policies 2022: Morocco (Country Profile), 2022). There are many small and medium enterprises, informal operators, and community tourism initiatives throughout Morocco, especially in regions rich with historic heritage. These operations create relationships with their customers and rely heavily on building trust in service delivery (Lakhlifi and Boumaize 2025). The introduction of online booking systems and chatbots, and the implementation of digital

Fig. 3 Conceptual schema of tourism management and AI adoption in Morocco. *Source* Authors' own elaboration based on Wilson and Chikhi (2020), Ez-Zahrany et al. (2024), Lakhlifi and Boumaize (2025), El Kaddouri and Benahmed (2022)

marketing by many tourism businesses has been developed, but few businesses have taken advantage of advanced artificial intelligence tools because of a lack of knowledge about using these tools, the urban–rural gap between access to technology, and cultural authenticity concerns (Zarrouk and El Aidouni 2023). As can be seen from the perspective of history, institutions, and culture; all three of these categories contribute to how businesses in Morocco adopt AI and that most do so largely to enhance their current business interactions and traditional methods of service delivery (UNESCO 2021).

4 AI in the Spanish Tourism Management

The Spanish tourism management environment is shaped by its institutionalization of a technology-based infrastructure that supports its national and regional government systems through smart tourism frameworks using a top-down approach for both strategic coordination and innovation in operations. The "Smart Tourism Destination (DTI)" model explicitly states that a tech-based component (i.e., ICTs, AI, and big data analytics) should be a part of each destination's management systems and processes, which are used as tools for improving decision-making, competitive advantage, and the delivery of services that are more customized for each individual client (Andrades and Martínez Marín 2024). AI and its usage within the DTI framework, have been found to help improve tourism intelligence and operational efficiencies. Examples of how AI will be used include: (1) Predictive Analytics to forecast demand, (2) Dynamic Pricing Models and (3) Recommendation Systems (Florido Benítez and del Alcázar Martínez 2024). As formalized management structures and performance goals exist in Spain, they create an environment with higher tolerances for the use of technology in the tourism environment, and they also enable the increasing integration of multi-language AI services to enhance the visitor experience across multiple markets. This also raises ethical issues related to data privacy and algorithmic bias—particularly in line with the EU's General Data Protection Regulations (GDPR), and the increased demand for upskilling of the workforce in order to successfully apply AI-based tools in an increasingly digital world (Hernández White and Rodríguez Díaz 2024) (Fig. 4).

The figure represents how the structure of Spain's decentralized and entrepreneurial government is the basis for its tourism sector (i.e. how it provides support for the development of new technologies) and how Spain is strategically utilizing Artificial Intelligence within its Tourism System rather than as a peripheral function (Mintur 2025).

As an example of the inclusion of AI & Big Data in Destination Management, Spain has built Smart Tourism Platforms using AI and Big Data that provide Destinations with the ability to monitor and support real-time decision making through the use of AI algorithms, while also providing destination and/or tourism managers with the same functionality via an online database. In addition to the above, Spain's Advanced Applications of AI include Dynamic Pricing and Predictive Analytics,

Fig. 4 Conceptual schema of tourism management and AI in Spain. *Source* Authors' own elaboration based on OECD (2025), Soler Alvarez et al. (2025), Tymoshchuk (2023), Commission (2023)

both of which are integrated to a formalized management logic and performance orientation; thus, demonstrating that the cultural acceptance of Technology as a tool to facilitate Efficiency and Competitiveness is inherent in Spanish Tourism Culture (Miró and García 2024).

However, the inclusion of the above technologies presents certain managerial challenges, such as the need to train (up-skill) human resources, identify and account for Ethical issues related to Algorithmic Bias and Data Privacy and find sustainable solutions to Tourism Overcrowding (Wang et al. 2025), Therefore, from Spain's experience it is evident that AI Adoption in Tourism is not merely a Technology Upgrade, but is a Culturally and Institutionally Mediated Process that functions in concert with the Institutional Frameworks of Governance, the Managerial Capabilities of the Organization, and the Ethical Imperatives of Society (Andrades and Martínez Marín 2024).

5 Comparative Analysis of Morocco and Spain in Tourism AI Integration

National cultures and management structures shape both how businesses adopt and use Artificial Intelligence (AI) and how people experience tourism. In Morocco, tourism management remains focused on being human-centric, relationship-based and based on high context. As such, Moroccan businesses are using AI cautiously and in a supportive manner to maintain the authenticity of their products, while simultaneously emphasizing the importance of trust and relationship building in service delivery (El Boukhari et al. 2025). By contrast, the Spanish tourism industry is much more systematic, process-orientated and operates within a lower context. As such, the structures of the Spanish tourism industry are set up for use and enable AI to provide support and structure as part of their operational systems to help companies enhance efficiency, expand into new markets and enhance their forecasting abilities (Sánchez et al. 2025) (Table 1).

In Morocco, the combination of collectivist cultural values and strong power-distance fosters a preference for a human-centered approach to AI implementation, relying on managerial cooperation and incremental innovations to produce tourism-related outputs that are sustainable, authentic, and trustworthy (Martin Lucas et al. 2025). In Spain, the emphasis on individualistic cultural values, a reduced power distance, and an innovation-oriented norm has resulted in the implementation of AI from a performance-based and automation-based standpoint using strategic and decentralized managerial processes that allow for increased efficiency, broader

Table 1 Country-specific pathways of AI adoption in tourism management: a comparative conceptual schema (Morocco versus Spain)

Morocco	Spain
Collectivism High power distance Strong human interaction Higher uncertainty avoidance	Higher individualism Lower power distance Innovation-oriented norms Lower uncertainty avoidance
AI adoption Low–moderate automation Human-in-the-loop systems Decision-support	AI adoption High automation intensity Institutionalized AI Predictive and
Service authenticity Trust and social legitimacy Social sustainability	Operational efficiency Cost reduction Personalization at scale Global competitiveness
Hierarchical governance Consensus-based decisions	Decentralized decision-making Performance orientation AI as productivity driver

Source Authors' own elaboration based on Gretzel et al. (2023), Huang and Rust (2025), Martin Lucas et al. (2025), Shin et al. (2025), Buhalis and Sinarta (2023)

personalization of service, and improved competitiveness in the international tourism sphere (Shin et al. 2025; Gretzel et al. 2023).

Further, the cultural and managerial environments greatly impact how extensively AI technology will be utilized and, therefore, demonstrate that different countries have unique requirements when implementing AI technologies into their tourism sector (Kalhoro 2025).

6 Managerial and Policy Implications

Through the comparative analysis of the ways AI adoption in Morocco and Spain is developed, we have found that cultural, governance, and managerial logic are the primary factors influencing how AI becomes integrated into the sector of tourism. Developing recommendations from these findings necessitates that the technology strategy aligns with cultural expectations and institutional capacities to effectuate ethical, inclusive and sustainable strategies for deploying AI (Shin et al. 2025).

6.1 For Tourism Managers

Tourism companies need to create culturally sensitive AI solutions that align with their destination's service expectations. The emphasis of the Moroccan Tourism Industry is on the importance of building relational trust between people rather than machines; therefore, AI technology should assist rather than replace human employees and should reinforce rather than diminish trust within the customer service experience (ChenG et al. 2023) For instance, in Morocco, an AI Chatbot could augment time-sensitive information distribution while allowing customers to still access the personalized services of a human agent. This bears out with regards to high-context social dynamics in Morocco (Ivanov et al. 2024).

Tourism companies in Spain have more opportunity to leverage AI technologies to maximize profitability compared to their Moroccan counterparts. Managers in Spain can optimize how they incorporate AI through strategic integrative approaches that include a combination of predictive analytics and human oversight in developing hybrid decision-making workflows, incorporating both data-driven insights and culturally-raised subtleties into the delivery of services (Li et al. 2025). The Comparative research has found that AI augmentation strategies that use AI to enable rather than replace employees provide greater employee satisfaction and enhance consumer perceptions when they are guided by a clear division of roles and training (Lu and Gursoy 2025).

6.2 For Policymakers

For Policymakers from Both Regions: To Ensure an Adequate Workforce, Both Policymakers Must Create Training Programs That Are Consistent with the Rapid Advancements in Digital Technology (AI) and Have the Appropriate Training Required for Employees of Both Regions for the Local Job Markets. These Programs Should Teach Employees Both the Technical Aspects of Digital Technology (AI) and the Soft Skills Necessary to Prepare Them for Successful Additions to the Work Forces of Digital Technology (AI) (Tian 2025). In Morocco, Support for Small and Medium-Sized Enterprises (SMEs) and Informal Tourism Enterprises is Important in Building Capacity and Capability to Support the Transition to the Digital Economy and Support the Bridging of Rural and Urban Locations Through Digital Technology (Al Khattab and Al Thawadi 2024). In Spain, Policymakers Have an Opportunity to Develop Ethical Guidelines on the Use of AI in Tourism That Include the EU Data Privacy Regulations and Algorithmic Equality Provisions and Develop Transparent, Accountable Systems That Will Be Viewed Favorably by Local Communities. Therefore, Policymakers in Both Regions Should Prioritize Developing Training Programs That Teach Digital Literacy and Soft Skills Necessary to Successful Transition to Digital Employment in Rural, Less Developed Areas (Van der Heide 2025). Therefore, Policymakers in Both Regions Should Prioritize Developing Training Programs That Teach Digital Literacy and Soft Skills Necessary to Successful Transition to Digital Employment in Rural, Less Developed Areas (Zhao et al. 2023).

6.3 For SMEs

Tourism SMEs need to implement AI using a phased approach at a measured pace to enable them to achieve their objectives while considering their service culture and resources. Since Moroccan SMEs have low levels of digitalization and attach great importance to being authentic in their offerings, a starting point should be to implement a technology solution with CRM to enhance customer service and use AI to perform the rudimentary tasks associated with routine activities. Doing this will free up time for these SMEs to offer more personalized forms of customer interaction (Martínez Caro and Martínez García 2024; Gretzel et al. 2023).

Spanish SMEs are situated in a more structured technological environment and will have the ability to leverage cloud-based AI solutions for managing their revenue and dynamically pricing goods and services, but again, they will be able to do so through incremental increases in their digital competencies as well (Gretzel et al. 2023).

7 Conclusion

The present study demonstrates that AI applied to tourism is best understood through contextual conditions as a socio-cultural and managerial process rather than merely through technology and/or economics. The proposed framework (the relationship between culture, manager orientation and AI use in tourism, how culture and managerial styles of governance structure the intensity of culture/the purpose of AI use in tourism systems) sets the stage for understanding how the respective cultures of Morocco and Spain represent very different institutional logics regarding how AI is used within tourism: in Morocco, AI is mainly viewed as a tool to support (preserve) human interaction and authenticity; in Spain, structured management governance supports performance-driven systems of governance (scale, efficiency and ability). Based on the findings, this paper will enhance tourism management research by providing pathways (rather than simply discussions) of how AI becomes institutionalized based on context. Additionally, the implications of this framework provide evidence that when managerial values and governance are not aligned with how AI is designed to deliver services, there is a negative impact on trust, acceptance and visitor experience. This is especially evident in tourism where relationships are integral. However, when managerial values and governance are aligned with the goals of adopting AI, tourism destinations become more competitive and are able to meet their sustainability objectives. From this study's context of AI in Tourism, a path is outlined for future scholarship to be developed to test the suggested propositions against multiple tourism locations and organizational structures. Future research should employ a longitudinal and mixed-methods approach to continue exploring the transition of cultural standards along with the transition of technology. Ultimately, this paper demonstrates that culture is one of the most significant factors influencing the amount of AI that is adopted in the tourism industry as well as the overall effect of AI on the tourism experience, thus emphasizing that there is a need for culturally sensitive and ethically sound AI practices in developing Global Tourism.

References

Abdullah Alhelal, A., Abdulaziz Alshiha, A., Samir Al-Romeedy, B.: AI driven sustainable competitive advantage in tourism and hospitality: mediating roles of digital culture and skills. Sustainability **17**(19), 8903 (2025). https://doi.org/10.3390/su17198903

Al Khattab, A., Al Thawadi, A.: Digital transformation and skills development in tourism SMEs in emerging markets: a case of local tourism adoption. Tour. Econ. **30**(4), 524–544 (2024). https://doi.org/10.1177/13548166231109775

Andrades, L., Martínez Marín, E.: The Spanish Model for Smart Tourism Destination Management, pp. 87–126. Springer (2024). https://doi.org/10.1007/978-3-031-60709-7_6

Buhalis, D., Leung, D.: Smart tourism destinations: governance, innovation, and technology. Tour. Manag. Perspect. **25**, 1–4 (2018). https://doi.org/10.1016/j.tmp.2017.11.001

Buhalis, D., Sinarta, Y.: Real-time co-creation and nowness service: lessons from tourism and hospitality. J. Travel Tour. Mark. **36**(5), 563–582 (2019). https://doi.org/10.1080/10548408.2019.1592059

Buhalis, D., Sinarta, Y.: Real-time co-creation and smart tourism ecosystems: towards a tourism digital ecosystem framework. J. Travel Tour. Mark. **40**(2), 197–212 (2023). https://doi.org/10.1080/10548408.2023.2179957

ChenG, M., Edwards, D., Goss, M.: Contextualising AI integration in hospitality: the role of service culture and employee augmentation. Int. J. Hosp. Manag. **113**, 9 (2023). https://doi.org/10.1016/j.ijhm.2022.103335

Chi, O.H.G., Gursoy, D., Nunkoo, R.: Customers' acceptance of artificially intelligent service robots: the influence of trust and culture. Int. J. Inf. Manag. **70**, 102623 (2023)

Commission, E.: Artificial Intelligence in Tourism: Opportunities, Challenges and Governance. EU Publications (2023)

Dwivedi, Y., Hughes, L., Ismagilova, E., et al.: Artificial intelligence (AI): multidisciplinary perspectives on emerging challenges, opportunities, and agenda for research. Int. J. Inf. Manag. **57**, 101994 (2021). https://doi.org/10.1016/j.ijinfomgt.2019.08.002

El Boukhari, M., Amajid, G., Fadili, M.: Tourism in the age of artificial intelligence: a conceptual perspective on marketing and beyond. Int. J. Digit. Appl. Manag. **4**(1), 16 (2025). https://doi.org/10.23882/ijdam.25227

El Kaddouri, A., Benahmed, N.: La digitalisation au service des touristes: la quête d'un avantage collaboratif. Revue Marocaine de Recherche en Management et Marketing **14**(2), 37967 (2022). https://doi.org/10.48376/IMIST.PRSM/remarem-v14i2.37967

Ez-Zahrany, C., Elmadhi, A., Mokharij, K.: L'Impact des technologies numériques sur le tourisme culturel et la valorisation du patrimoine au Maroc. Revue du LaRSLAM **12**(10), 12 (2024). https://doi.org/10.34874/IMIST.PRSM/larslam-i12.49515

Florido Benítez, L., del Alcázar Martínez, B.: How artificial intelligence (AI) is powering new tourism marketing and the future agenda for smart tourist destinations. Electronics **13**(21), 8–9 (2024). https://doi.org/10.3390/electronics13214151

Ghesh, N., Alexander, M., Davis, A.: The artificial intelligence-enabled customer experience in tourism: a systematic literature review. Tour. Rev. **79**(1), 1017–1037 (2023). https://doi.org/10.1108/TR-04-2023-0255

Gretzel, U., Sigala, M., Xiang, Z., Koo, C.: Smart tourism challenges and opportunities in the post-digital era. Electron. Mark. **33**(1), 1–17 (2023). https://doi.org/10.1007/s12525-022-00590-1

Henriques, H., Pereira, L.N.: Hotel demand forecasting models and methods using artificial intelligence: a systematic literature review. Tour. Manag. Stud. **20**(2183), 21 (2024). https://doi.org/10.18089/tms.20240304

Hernández White, C., Rodríguez Díaz, B.: Tourism and ICTs: Advances in Data Science, Artificial Intelligence and Sustainability, pp. 4–6. Springer (2024)

Huang, M., Rust, R.: Artificial intelligence in service. J. Serv. Res. **21**(2), 155–172 (2018). https://doi.org/10.1177/1094670517752459

Huang, M., Rust, R.: Generative artificial intelligence in tourism management: an integrative review and roadmap for future research. Tour. Manage. **110**, 35 (2025). https://doi.org/10.1016/j.tourman.2025.105179

Ivanov, S., Webster, C., Berezina, K.: Adoption of robots and service automation in tourism. Inf. Technol. Tour. **26**(2), 131–148 (2024). https://doi.org/10.1007/s40558-024-00258-6

Jan, J., Khaled, A., Peggy, L.: Hofstede's cultural dimensions in technology acceptance models: a meta analysis. Univers. Access Inf. Soc. **23**, 717–741 (2024). https://doi.org/10.1007/s10209-022-00930-7

Kalhoro, M.A.-N.: Artificial intelligence and tourism 4.0 for sustainable transformation: a systematic review of environmental and governance impacts. Environ. Energy Res. **2**(4), 58 (2025). https://doi.org/10.36922/EER025320058

Keding, C.: Understanding the interplay of artificial intelligence and strategic management: four decades of research in review. Manag. Rev. Q. **71**(1), 91–134 (2021). https://doi.org/10.1007/s11301-020-00181-xideas.repec.org
Lakhlifi, H., Boumaize, A.: Formulation of Morocco's digital tourism policy: empirical evidence from ARDL bounds testing. Int. J. Res. Econ. Financ. **2**(8), 30–50 (2025). https://doi.org/10.71420/ijref.v2i8.142
Li, H., Xi, J., Cathy, H.C., Bruce, X.B., Zheng, X.: Generative artificial intelligence in tourism management: an integrative review and roadmap for future research. Tour. Manag. **110**, 105179 (2025). https://doi.org/10.1016/j.tourman.2025.105179
Li, X., Zhang, J., Wang, Y.: Hybrid human–AI decision support systems in tourism: strategic implications and governance. J. Travel Res. **64**(1), 15–32 (2025). https://doi.org/10.1177/00472875221098765
López-Naranjo, A.L., Puente-Riofrio, M.I., Carrasco-Salazar, V.: Artificial intelligence in the tourism business: a systematic review. Front. Artif. Intell. **8**, 12–16 (2025). https://doi.org/10.3389/frai.2025.1599391
Lu, L., Gursoy, D.: Drivers and barriers to AI adoption in hospitality and tourism SMEs: an empirical study. Tour. Manage. **97**, 104815 (2025). https://doi.org/10.1016/j.tourman.2024.104815
Martin Lucas, M., Medina-Merodio, J., Robina-Ramirez, R.: What role does the humanisation of artificial intelligence play in improving tourist performance? Sociología y Tecnociencia **15**(1), 223–248 (2025). https://doi.org/10.24197/st.1.2025.223-248
Martínez Caro, E., Martínez García, R.: Aligning AI strategy with SME capacity in European destinations. J. Destin. Mark. Manag. **28**, 25 (2024). https://doi.org/10.1016/j.jdmm.2024.100802
Mengying, Z., Furui, Y., Dogan, G.: The effects of generative artificial intelligence on consumers in hospitality and tourism: a systematic review and future research directions. Int. J. Hosp. Manag. **133**, 3–4 (2026)
Mintur: AI in Tourism Plan. Récupéré sur Spanish Ministry of Industry and Tourism. https://www.mintur.gob.es/es-es/GabinetePrensa/NotasPrensa/2025/Paginas/Hereu-anuncia-desarrollo-de-un-plan-integrado-para-el-uso-de-Inteligencia-Artificial-Tur%C3%ADstica-en-Espa%C3%B1a.aspx (2025)
Miró, A., García, J.L.: Artificial intelligence and its importance in the near future in the Spanish tourism industry. PASOS Revista de Turismo y Patrimonio Cultural, 2–8 (2024)
OECD: OECD Tourism Trends and Policies 2022: Morocco (Country Profile). OECD Publishing. www.oecd.org/en/publications/oecd-tourism-trends-and-policies-2022_a8dd3019-en/full-report/morocco_2e779207.html? (2022)
OECD: Récupéré sur SEGITTUR. Data spaces and innovation across the tourism sector in Spain. Organisation for Economic Co operation and Development. https://www.oecd.org/content/dam/oecd/en/publications/reports/2025/07/strengthening-the-tourism-statist (2025)
Sánchez, S., Malonda, J., González, M.: Inteligencia artificial y transformación estructural del turismo español: hacia una inteligencia turística aumentada basada en infraestructuras públicas digitales y dinamización empresarial. Estudios Turísticos **230**, 8–12 (2025). https://doi.org/10.61520/et.2302025.1258
Sánchez-Martín, J., Guillén-Peñafiel, R., Hernàndez-Carretero, A.-M.: Artificial intelligence in heritage tourism: innovation, accessibility, and sustainability in the digital age. Heritage **8**(10), 428 (2025). https://doi.org/10.3390/heritage8100428
Sharma, A.: Artificial intelligence-driven predictive analytics in tourism management. J. Tour. Futur. **10**(2), 145–159 (2024). https://doi.org/10.1108/JTF-03-2023-0064
Shin, H., Ryu, J., Jo, Y.: Navigating artificial intelligence adoption in hospitality and tourism: managerial insights, workforce transformation, and a future research agenda. Int. J. Hosp. Manag. **128**(104187), 15–18 (2025). https://doi.org/10.1016/j.ijhm.2025.104187
Soler Alvarez, B., Anaya Sánchez, R., Rossi Jimenez, C.: Aplicación de la inteligencia artificial en turismo: una revisión sistemática de la literatura. Estudios Turísticos **230**, 1259 (2025). https://doi.org/10.61520/et.2302025.1259

Sousa, A., Cardoso, P., Dias, F.: The use of artificial intelligence systems in tourism and hospitality: the tourists' perspective. Adm. Sci. **14**(8), 165 (2024). https://doi.org/10.3390/admsci14080165
Tian, F.: Policy frameworks for AI and tourism workforce transformation: a global perspective. Tour. Manag. Perspect. **40**, 12 (2025). https://doi.org/10.1016/j.tmp.2024.100988
Tussyadiah, I.: A review of research into automation in tourism: launching the annals of tourism research curated collection on artificial intelligence and robotics in tourism. Ann. Tour. Res. **81**, 1–14 (2020). https://doi.org/10.1016/j.annals.2020.102883
Tymoshchuk, O.: The forms of digitalization of the tourism industry of Spain. Business Inform. **4**(15), 15–20 (2023). https://doi.org/10.32983/2222-4459-2023-4-15-20
UNESCO. https://www.unesco.org/ethics-ai/en/morocco?utm (2021)
Van der Heide, E.J.: Governing ethical AI use in public services: lessons for tourism governance. Gov. Inf. Q. **42**(1), 35–39 (2025). https://doi.org/10.1016/j.giq.2024.101712
Viglia, G., Pera, R., Bigné, E.: The determinants of stakeholder engagement in digital platforms: a service ecosystem perspective. J. Bus. Res. **89**, 404–414 (2018). https://doi.org/10.1016/j.jbusres.2017.12.036
Wang, S., Wang, Q., Cui, Q., Lan, T.: Artificial intelligence in tourism: a systematic literature review and future research agenda. Sustainability **17**(20), 5–16 (2025). https://doi.org/10.3390/su17209080
Wilson, D., Chikhi, S.: Tourism micro-entrepreneurship and informal employment in North Africa: patterns, prospects and policies. Tour. Econ. **26**(8), 1381–1404 (2020). https://doi.org/10.1177/135481661986695
Yam, K., Tan, T., Jackson, J., Shariff, A.: Cultural differences in reactions to robots, algorithms, and artificial intelligence. Manag. Organ. Rev. **25**, 15 (2023). https://doi.org/10.1017/mor.2023.15
Zarrouk, I., El Aidouni, A.: The digital marketing practiced by the Moroccan tourism sector: national-scale communication actions and environmental management. E3S Web Conf. **412**, 41 (2023). https://doi.org/10.1051/e3sconf/2023412010
Zhao, Y., Feng, J., Law, R.: Ethical design of tourism recommender systems: balancing personalization with privacy. Tour. Manag. **94**, 104575 (2023). https://doi.org/10.1016/j.tourman.2023.104575

Brand Hate Detection Using Natural Language Processing Development and Illustration Case

Mohamed Assoud and Amira Sebbah

Abstract This paper proposes and illustrates an explainable Natural Language Processing (NLP) pipeline for detecting and classifying brand hate in online consumer reviews. Moving beyond traditional sentiment analysis approaches that reduce consumer feedback to simple positive–negative polarity, the study adopts a hybrid lexicon-based framework that combines sentiment scoring, emotion profiling using the NRC Emotion Lexicon, and rule-based classification to operationalize the multidimensional nature of brand hate. The pipeline is implemented as an interactive R Shiny application and follows a structured workflow encompassing automated review scraping from ConsumerAffairs, sentiment-based filtering of negative content, textual preprocessing, emotion analysis, and a two-stage classification procedure distinguishing Mild, Moderate, Strong, and Hybrid forms of brand hate. To enhance transparency and interpretability, analytical outputs are communicated through interactive visualizations, including hate intensity distributions, reason-based breakdowns, emotion word clouds, and sentiment–intensity bubble plots. The artifact is demonstrated through a step-by-step application using real consumer reviews, serving to document system behavior and analytical logic rather than to test theory or generalize empirical findings. By translating established brand hate theories into a computationally accessible and explainable tool, this study contributes a methodological foundation for future large-scale, longitudinal, and cross-platform investigations of consumer hostility, while offering managers a practical means to monitor and understand emerging brand hate dynamics.

Keywords Brand hate · Natural language processing · Emotion analysis · Artificial intelligence · Consumer reviews

M. Assoud (✉)
ISGA Rabat, Rabat, Morocco
e-mail: m.assoud@uhp.ac.ma

M. Assoud · A. Sebbah
Hassan First University of Settat, Settat, Morocco

A. Bouziane and K. Bouziane (eds.), *AI, Management, and Culture: Shaping Organizations in the Digital Age*, Studies in Computational Intelligence 1291,
https://doi.org/10.1007/978-3-032-27882-1_2

1 Introduction

In the current digital landscape, consumer voice has become more than opinion—it is power (Kucuk 2020). Online reviews now guide the most purchase decisions, with 89% of global consumers consulting them before buying (Zhang et al. 2025). These reviews are not just signals of satisfaction or frustration; they shape reputations, shift market dynamics, and can directly influence sales (Assoud and Berbou 2026a). A single review, especially if negative, can reach thousands, alter brand perception, and even boost or kill conversion rates (Lis and Fischer 2020). This shift marks a structural transformation in brand-consumer relations. Thanks to platforms like Amazon, TripAdvisor and ConsumerAffairs, consumers now have tools to openly challenge, criticize, or even mobilize against brands (Loureiro and Kaufmann 2018; Parveen 2025). What makes this even more impactful is the psychological negativity bias—our tendency to process and retain negative experiences more intensely than positive ones (Ito et al. 1998; Zhang et al. 2025). In review environments, this means negative content is not only more visible but more trusted and persuasive. In effect, digital negativity travels faster and cuts deeper (Kucuk 2020).

Within this context, brand hate has emerged as a central concept to explain how dissatisfaction escalates into public hostility (Assoud and Berbou 2025a; Fetscherin 2019; Zhang and Laroche 2020). It is not simply about being let down. Brand hate reflects a deeper emotional rejection—a rejection that is often moral, symbolic, or personal (Hashim and Kasana 2019; Kucuk 2019a; Zarantonello et al. 2018). Consumers don't just stop buying; they retaliate, warn others, and in some cases, attack the brand narrative itself. Prior studies have linked this emotion to symbolic incongruity, ideological incompatibility, ethical violations, or negative experiences (Assoud and Berbou 2026b; Costa and Azevedo 2024; Sameeni et al. 2024; Santos et al. 2024). On emotional level, Brand hate is now recognized as multidimensional (Kucuk 2019a, b). Zhang and Laroche (2020) identify primary emotions such as anger, sadness, fear, disgust and surprise that form the emotional foundation. Fetscherin (2019) proposes a typology of hate forms, while Sternberg and Sternberg (2008) offer a psychological framework—the triangular theory of hate—that explains how disgust, anger, and commitment to rejection combine to sustain hostility. Together, these models help us understand not just why consumers hate but how intensely, and through what emotional lens.

Despite its conceptual richness, brand hate remains analytically underdeveloped (Assoud and Berbou 2026b). Two main gaps stand out—one methodological, the other managerial. First, brand hate research has largely relied on primary data collection—structured surveys (Assoud and Berbou 2025a; Fetscherin 2019; Sameeni et al. 2024; Santos et al. 2024), interviews (Bryson et al. 2013; Sharma et al. 2022; Valenzuela et al. 2022), and experiments (Costa and Azevedo 2024; Yang and Mundel 2022). These tools have helped define the concept, understand its antecedents and outcomes, but they fall short when it comes to capturing how brand hate really unfolds: online, in real time, in raw, emotional language. What is striking is that while Natural Language Processing, machine learning and deep learning have become mainstream in business research—used for predicting churn (Kang et al. 2020), analyzing sentiment (Albladi et al. 2025), and modeling brand equity (Yu and Chauhan 2025)—they have been largely absent in brand hate studies. This is a critical oversight. Theories are evolving, but the methods are lagging. Without computational tools, we miss the intensity, emotion, and nuance that define hate in the digital sphere. On the practitioner side, the challenge is access. Existing sentiment analysis tools often reduce complex emotions to simple polarity: positive vs. negative. They don't distinguish between mild dissatisfaction and deep-seated hate. Worse, when AI is used—such as in transformer models like GPT-2 (Mednini et al. 2024) —the systems often behave like black boxes. They detect something, but they do not explain it. And for a brand manager trying to understand risk or craft a response, opacity is not an option. These tools are also expensive and resource-intensive, limiting their adoption among small and medium businesses. In short, while brands are exposed to rising digital hostility, they still lack usable, affordable, and explainable tools to detect, understand, and manage it (Assoud and Berbou 2025b).

Against this background, the current paper aims to address the following question: How can natural language processing (NLP) be leveraged to detect and classify brand hate in online consumer reviews, while providing actionable insights for managers? More specifically, we intent to (1) Develop a multidimensional brand hate classification system, grounded in psychological models and adapted for computational analysis (Assoud and Berbou 2025a, b; Fetscherin 2019; Kucuk 2019a; Sternberg and Sternberg 2008; Zhang and Laroche 2020) (2) Design an accessible and interpretable NLP tool that goes beyond simple polarity detection, identifying the intensity, emotions, and justifications underlying brand hate; and (3) Apply and validate this tool on real consumer reviews, aiming to address the limitations of interpretability and accessibility identified in previous research.

2 Literature Review

2.1 *Brand Hate*

Brand hate represents one of the most intense and consequential forms of negative consumer-brand relationships, emerging as a critical area of inquiry within consumer behavior and marketing research over the past decade (Yadav and Chakrabarti 2022). Unlike simple dissatisfaction or dislike, brand hate encompasses deep-seated emotional hostility that fundamentally alters the nature of consumer-brand interactions and can have profound implications for brand performance, market dynamics, and consumer well-being (Assoud and Berbou 2025b). The phenomenon has gained increasing attention from both academic researchers and marketing practitioners as digital platforms have amplified consumers' ability to express, share, and mobilize negative emotions toward brands on an unprecedented scale (Kucuk 2020).

The conceptual foundations of brand hate research can be traced to broader theoretical developments in consumer-brand relationships (Fournier 1998), which have traditionally focused on positive emotions such as brand love (Carroll and Ahuvia 2006), attachment (Arvidsson 2006), and loyalty (Curtis et al. 2011). However, the recognition that consumer emotions exist on a continuum that includes both positive and negative extremes has led researchers to explore the darker dimensions of brand relationships. This shift reflects a growing understanding that negative emotions toward brands are not merely the absence of positive feelings but constitute distinct psychological states with unique antecedents, characteristics, and consequences that require specialized theoretical and empirical attention (Romani et al. 2012; Zarantonello et al. 2016).

2.1.1 Conceptualization and Emotional Components

Conceptually, brand hate represents an intense negative emotional state that consumers develop toward specific brands, characterized by sustained hostility and aversion. Kucuk (2019a) defines brand hate as "a psychological state in which a consumer forms an intense negative emotion and hatred toward a brand, a hatred that reveals itself in feelings of anger, contempt, and disgust." This definition establishes brand hate as a distinct psychological phenomenon that goes beyond simple dissatisfaction or dislike, encompassing deep-seated emotional hostility that fundamentally alters consumer-brand relationships. Unlike temporary negative emotions, brand hate is characterized by its persistence, intensity, and the active desire to distance oneself from or harm the hated brand.

The emotional structure of brand hate has been conceptualized through several theoretical frameworks that reveal its multidimensional nature. The triangular theory of hate, originally developed by Sternberg and Sternberg (2008) and adapted to brand contexts, identifies three core components: negation of intimacy (emotional distance and coldness), passion (intense negative emotions), and commitment (sustained negative feelings over time). Building on this foundation, Fetscherin (2019) proposed five distinct types of brand hate based on different combinations of these components: cool hate (negation of intimacy only), hot hate (passion only), cold hate (commitment only), boiling hate (negation of intimacy and passion), and burning hate (passion and commitment). Zhang and Laroche (2020) further refined our understanding by identifying five primary emotions (Shaver et al. 1987) underlying brand hate: anger, sadness, fear, surprise, and disgust. Their tripartite model organizes these emotions into three categories: anger-related emotions, sadness-related emotions, and fear-related emotions. This emotional framework demonstrates that brand hate is not a unitary construct but comprises multiple emotional dimensions that can manifest in different combinations depending on the triggering circumstances and individual consumer characteristics.

2.1.2 Antecedents and Behaviors

Research has consistently identified three primary categories of brand hate antecedents. First, negative Past Experience that represent the most direct pathway to brand hate, occurring when brands fail to meet consumer expectations through poor product performance, service failures, or perceived unfair treatment (Curina et al. 2020; Hegner et al. 2017; Pinto and Brandão 2021). These experiential failures create feelings of betrayal and injustice that can escalate into sustained hatred, particularly when consumers perceive the brand as having violated fundamental obligations or acted with malicious intent (Oliver 1980). Second, Symbolic Incongruity emerges when consumers perceive misalignment between a brand's image, values, or associations and their own self-concept or identity (Gois et al. 2022; Islam et al. 2019). This pathway to brand hate is particularly relevant in contexts where brands serve as vehicles for self-expression and identity construction. When consumers feel that a brand no longer represents or conflicts with their desired identity, they may develop hatred as a means of psychological distance and identity protection. Third, Ideological Incompatibility that occurs when brand values, actions, or positions conflict with consumer ethical, political, or social beliefs (Abbasi et al. 2023; Costa and Azevedo 2024). This form of brand hate often extends beyond individual experiences to encompass broader social and cultural conflicts, making it particularly volatile and difficult to manage. The rise of brand activism and corporate social responsibility has increased the salience of ideological factors, as brands increasingly engage with social and political issues that may alienate certain consumer segments.

Brand hate manifests through two primary behavioral pathways that reflect the fundamental fight-or-flight response to perceived threats (Bayarassou et al. 2020; Hirschman 1970). Flight behaviors represent passive responses aimed at minimizing contact with the hated brand, including brand avoidance (Assoud and Berbou 2025a; Xin Ying Ong and Vila-Lopez 2024), patronage reduction (Brandão et al. 2023; Odoom et al. 2024), and quiet switching to competitor brands (Fetscherin 2019). These withdrawal behaviors allow consumers to protect themselves from further negative experiences while expressing their dissatisfaction through market-based actions. Fight behaviors encompass active responses aimed at confronting or harming the hated brand, including negative word-of-mouth (Sajid et al. 2024; Sameeni et al. 2024), brand retaliation (Costa and Azevedo 2023; Fetscherin 2019), and participation in boycotts (Husnain et al. 2020) or anti-brand communities (Brandao and Popoli 2022). These aggressive responses are more likely when brand hate is characterized by high levels of anger and when consumers perceive that the brand has acted intentionally or maliciously. The digital transformation of consumer behavior has amplified the potential impact of fight responses, as consumers can now organize collective action and reach large audiences with their negative messages through social media platforms and online communities.

2.2 Natural Language Processing

Natural Language Processing (NLP) is a branch of artificial intelligence and computational linguistics that focuses on enabling computers to understand, interpret, generate, and respond to human language in a meaningful way (Manning and Schutze 1999). It integrates techniques from computer science, linguistics, and mathematics to facilitate communication between humans and machines through natural language. NLP comprises two primary tasks: Natural Language Understanding (NLU)—the process of interpreting human language for downstream analysis—and Natural Language Generation (NLG)—the automated creation of human-readable text based on data inputs (McDonald 2010). Recent advancements in neural networks and deep learning have significantly enhanced NLP capabilities, particularly in applications such as machine translation, sentiment analysis, speech recognition, and conversational agents (Albladi et al. 2025; Syauqi and Wibawa 2025).

Natural Language Processing (NLP) typically unfolds through a series of broad steps that allow researchers to transform unstructured text into meaningful insights. The process begins with preparing the text, which includes cleaning the data by removing unnecessary elements like symbols or duplicates and breaking down the text into individual words or phrases (Feuerriegel et al. 2025). The next step is to represent the text in a structured format that computers can understand—this often means identifying how frequently certain words appear or how similar different texts are (Albladi et al. 2025). Once the data is prepared and structured, researchers apply analytical techniques to detect patterns, topics, or emotions in the text, using models that can identify dominant themes, classify opinions, or assess sentiment (Albladi et al. 2025; Liu et al. 2012). Finally, the results are interpreted and visualized in ways that support decision-making—for example, by showing which issues dominate consumer complaints or which words are most associated with negative sentiment (Kang et al. 2020). Altogether, these steps allow scholars and managers to make sense of large volumes of language-based data in a reliable and scalable manner.

Researchers in business studies have increasingly adopted Natural Language Processing (NLP) to analyze vast amounts of unstructured textual data such as corporate reports, social media posts, and customer feedback. In accounting, NLP has been used to extract topics from 10-K filings and analyze how these disclosures affect investor perceptions; for instance, Dyer et al. (2017) applied topic modeling to examine the evolution of disclosure content and its market implications. In finance, Calomiris and Mamaysky (2019) analyzed global financial news using sentiment algorithms and entropy measures to understand risk dynamics and returns across markets. In information systems, Pan et al. (2019) utilized cosine similarity on business descriptions to assess how threats of new market entry affect R&D investment decisions. In strategic management, Barlow et al. (2019) applied NLP to Google Play app descriptions to measure prototype and exemplar similarity, thereby capturing strategic distinctiveness in market entry. Collectively, these studies illustrate how NLP enhances empirical rigor, objectivity, and scale in management research by transforming textual data into actionable insights.

In marketing research, NLP has become an essential tool for decoding consumer sentiment, refining product positioning, and managing brand reputation. Marketing scholars analyze online reviews, social media posts, and user-generated content using a range of techniques—from traditional approaches like part-of-speech tagging and bag-of-words to advanced models such as Convolutional Neural Networks (CNN) and Long Short-Term Memory (LSTM) networks (Kang et al. 2020). Sentiment analysis, both lexicon-based and algorithmic, is widely employed to detect emotional tone, while topic modeling is used to extract product features, innovation signals, or themes in advertising content (Yu and Chauhan 2025). Studies have also applied clustering methods and ensemble learning models to improve classification accuracy and interpretability (Parveen 2025). For example, Melumad et al. (2019) examined how the use of smartphones affects the emotional expression in consumer reviews, revealing that mobile-written reviews tend to be more emotionally intense. With the rise of social media and real-time feedback, NLP equips marketers with dynamic, data-driven tools to better understand market sentiment, personalize engagement, and respond to crises (Kang et al. 2020).

Despite the growing integration of NLP across management disciplines, brand hate research has largely remained anchored in traditional, manual methods such as survey-based measurement (Assoud and Berbou 2025a; Fetscherin 2019), manual content analysis (Zarantonello et al. 2018), or qualitative research (Bryson et al. 2013; Farhat and Chaney 2021). While these approaches have provided valuable conceptual insights into the antecedents and outcomes of brand hate, they often rely on small samples, static data, and subjective interpretation. This contrasts sharply with the computational capabilities of NLP, which enable the large-scale, real-time analysis of consumer-generated content such as reviews, complaints, and social media posts.

A notable attempt to bridge this methodological gap is the recent work of Mednini et al. (2024), who applied a fine-tuned GPT-2 model to detect brand hate speech across multilingual datasets (Arabic, French, and English). Their chatbot-based sentiment classification system achieved impressive accuracy rates—0.98 in training and 0.84 in testing—demonstrating the potential of transformer-based models for identifying hate-laden consumer language. However, despite these promising technical results, significant limitations remain. The GPT-2 model operates as a "black box," making it difficult for managers to interpret the reasoning behind the model's classifications, which reduces its usefulness for decision-making and corrective action planning. Additionally, the implementation of such transformer models entails considerable computational costs, technical expertise, and infrastructure, which may exceed the capacity of many firms, especially SMEs. Moreover, most existing NLP approaches focus solely on detecting the presence of negative sentiment or hate speech without providing a structured classification of brand hate based on its emotional components or underlying motivations. This lack of interpretability and diagnostic depth limits their utility for strategic brand management. Finally, because the model is trained on labeled data within a controlled research context, its generalizability to noisy, fast-evolving real-world platforms remains uncertain. These challenges underscore the need for more accessible, explainable, and cost-effective NLP tools that can provide actionable insights to managers seeking to proactively monitor and address brand hate in increasingly complex digital environments.

Building on these insights and addressing the identified gaps, the next section outlines the methodological approach adopted to develop and validate our brand hate detection tool.

3 Research Methodology

3.1 *Programming Language and Technical Framework*

The brand hate detection tool was developed using the R programming language in combination with the Shiny web application framework. R is a widely recognized statistical computing environment, originally designed for data analysis and visualization, and it continues to be the tool of choice for applied statistics, social science research, and quantitative modeling (Dege and Brüggemann 2024). It was selected for this project due to its robust capabilities in natural language processing (NLP), sentiment analysis, and data visualization, which are essential for emotion-based consumer review analysis (Yildirim and Kübler 2023). The tidytext, sentimentr, and syuzhet packages, in particular, offer pre-integrated access to widely used sentiment lexicons—such as the NRC Emotion Lexicon—supporting transparent and theoretically grounded emotion detection (Mohammad and Turney 2013). As an open-source tool, R ensures transparency, reproducibility, and accessibility, making it an ideal choice for academic research and open innovation. The Shiny framework enables the deployment of interactive dashboards without requiring frontend development expertise, allowing users to run real-time text scraping, emotion classification, and visualization in a single interface (Doi et al. 2016). This lightweight, server-efficient approach contrasts with more complex machine learning pipelines, which often require labeled data and computational infrastructure. Moreover, R integrates smoothly with web scraping packages such as rvest, making it possible to retrieve live consumer reviews (Wickham and Bryan 2023). On these basis, our Shiny application offers a low-code, real-time, and explainable solution for detecting brand hate intensity directly from user-generated content—bridging academic insights with actionable brand monitoring.

3.2 *Data Base*

ConsumerAffairs was selected as the primary data source for this study due to its credibility, transparency, and extensive coverage of consumer experiences across multiple industries. Unlike platforms such as Amazon, which primarily focuses on product-specific reviews, or TripAdvisor, which centers on travel, tourism, and hospitality,

ConsumerAffairs offers a broad, cross-sectoral perspective, encompassing industries such as technology, telecommunications, finance, healthcare, food services, and personal care. This diverse industry representation provides a richer basis for analyzing brand-related emotions, especially in the context of brand hate, which can manifest in both product and service experiences.

All reviews on ConsumerAffairs are written in English, which aligns well with the linguistic requirements of most established sentiment lexicons and NLP packages used in this study (Mohammad and Turney 2013). As such, no language filtering was necessary, and the dataset was fully compatible with tools used for emotion detection and brand hate classification. The consistent use of English eliminates potential issues of translation loss, misinterpretation, or lexicon mismatch, thereby enhancing the reliability and comparability of the textual analysis. Another key feature of ConsumerAffairs is that reviewers are required to justify their rating, regardless of whether it is positive or negative. This results in context-rich, narrative-style reviews that offer detailed explanations of consumer experiences and sentiments. These open-text fields are particularly suitable for natural language processing and enable the detection of emotionally charged language often associated with brand hate (Feuerriegel et al. 2025; Kang et al. 2020).

Moreover, ConsumerAffairs ensures data reliability and integrity by verifying user identities through contact information, employing AI-assisted moderation to detect anomalies or manipulative behavior, and engaging human moderators to assess the quality, relevance, and helpfulness of submitted reviews. As of 2025, the platform hosts over 4.7 million verified reviews, offering a large and methodologically sound sample of user-generated content for real-time, scalable analysis. Taken together, the language consistency, cross-industry diversity, and review depth make ConsumerAffairs a particularly suitable and robust source for conducting large-scale, emotion-sensitive brand hate research using natural language processing.

3.3 Pipeline Overview

The analytical pipeline is structured into four sequential stages and twelve steps, each designed to progressively transform raw consumer reviews into actionable brand hate insights. Stage 1 (Data Collection) ensures the integrity and contextual relevance of input by extracting brand metadata and scraping consumer reviews from multiple pages. Stage 2 (Data Preprocessing) filters out non-relevant content and prepares the textual data through sentiment-based filtering and tokenization. Stage 3 (Data Analysis) applies lexicon-based emotion detection and hybrid classification techniques to identify the type and intensity of brand hate. Finally, Stage 4 (Data Visualization and Reporting) synthesizes the findings through intuitive visual tools to facilitate interpretation, communication, and managerial application. Figure 1 illustrates the pipeline overview.

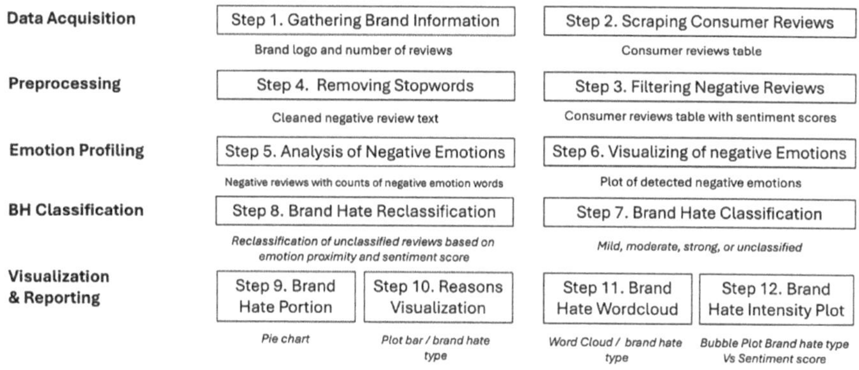

Fig. 1 Pipeline overview

4 Operational Demonstration: Stage-by-Stage Application

This section presents an operational demonstration of the proposed brand hate detection artifact through a stage-by-stage application. To enhance transparency and interpretability, the demonstration relies on KFC as an illustrative case, selected solely to exemplify how the system processes real-world consumer reviews. Importantly, KFC is not treated as an empirical case study nor as a basis for theory testing or generalization. Instead, it serves as a concrete illustration that allows each pipeline stage—from data collection to brand hate classification and visualization—to be examined in detail. The figures reported in this section document system behavior and analytical logic rather than empirical results, in line with design-science research conventions.

4.1 Data Collection

The first stage focuses on data collection, beginning with the extraction of brand metadata—name, review count, and logo—from a ConsumerAffairs URL to ensure contextual accuracy. It then scrapes and deduplicates consumer reviews across multiple pages, generating a structured dataset for subsequent emotion and brand hate analysis.

4.1.1 Brand Metadata

The analytical pipeline begins with the extraction of brand metadata from a ConsumerAffairs URL provided by the user. This step ensures that the correct brand is being analyzed and that sufficient data is available for subsequent processing. Specifically, the application retrieves three key elements directly from the webpage:

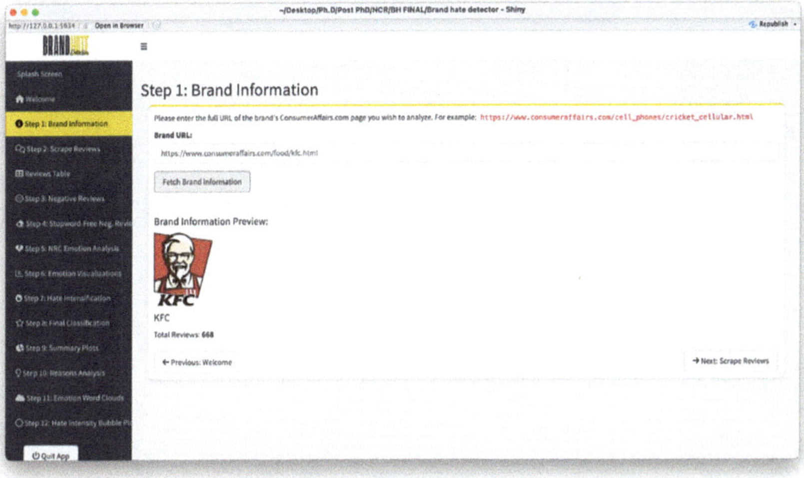

Fig. 2 KFC metadata and information

the brand name, the total number of available reviews, and the brand's logo image. The brand name is extracted from the page header, while the number of reviews is collected from the review count element. The logo is retrieved and rendered within the interface as a visual cue to help users verify that the page matches their intended target. This automated data extraction is implemented using the rvest package (Wickham and Bryan 2023), a widely used R tool designed for web scraping. rvest wraps around the xml2 and httr libraries, allowing the app to download and parse HTML documents using simple, readable syntax. This step establishes the integrity and relevance of the input data, creating a reliable foundation for emotion-based analysis in subsequent stages.

Figure 2 displays the automated extraction of brand metadata for the KFC illustration, including brand name, review volume (668), and visual identity. This output demonstrates the system's ability to verify contextual accuracy before initiating textual analysis, thereby reducing the risk of misclassification at later stages.

4.1.2 Scrape Consumer Reviews

Following brand identification, the application proceeds to scrape consumer reviews from the specified ConsumerAffairs page using the rvest package (Wickham and Bryan 2023). The scraping process retrieves visible review text from multiple paginated HTML pages, enabling the collection of a substantial volume of consumer feedback. To ensure data quality, all duplicate entries are automatically removed, and each review is assigned a unique identifier (review_id) for traceability. The

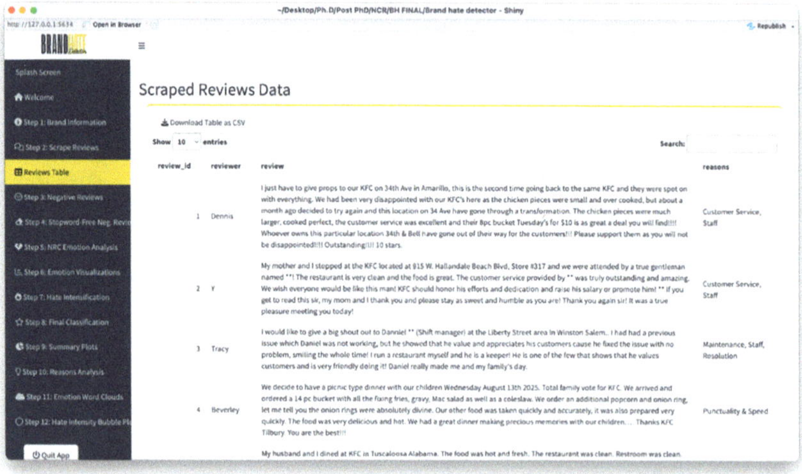

Fig. 3 Sample of scraped KFC consumer reviews from ConsumerAffairs

resulting dataset is displayed in a structured and searchable table that includes relevant metadata such as the reviewer's name, full review text, and the associated reason for complaint, where applicable. This step generates a clean, organized, and review-level dataset that serves as the foundation for subsequent emotional and brand hate analysis.

Figure 3 presents a sample of consumer reviews scraped from ConsumerAffairs for the KFC illustration. The figure demonstrates the artifact's ability to automatically retrieve, deduplicate, and structure review-level data, including review text and associated metadata, in a format suitable for subsequent sentiment and emotion analysis. This step establishes the empirical input for the pipeline while ensuring that downstream analyses operate on clean and contextually verified consumer-generated content.

With the dataset of consumer reviews systematically collected and structured, the next critical step involves preparing this raw textual data for robust Analysis. This preprocessing phase ensures that only relevant and analytically meaningful content is retained, thereby laying a solid foundation for subsequent emotion and brand hate detection.

4.2 Data Pre Processing

The second stage involves data preprocessing to enhance analytical precision. It begins by filtering out all neutral and positive entries using sentiment scores, retaining only negatively valanced reviews likely to contain brand hate. Subsequently, reviews are tokenized and stripped of stopwords to preserve only meaningful content words, thereby improving the accuracy of emotion classification and hate detection in the following stages.

4.2.1 Filter Negative Reviews

Once the full set of reviews is collected and cleaned. The pipeline proceeds to automatically identify and retain only negatively valanced reviews. This is achieved using the sentimentr package (Rinker 2017), which calculates sentence-level sentiment scores based on a context-sensitive, lexicon-based approach. Each review is assigned a numerical sentiment score that reflects its emotional polarity, with negative values indicating unfavorable sentiment. The package filters out all neutral or positive reviews, preserving only those whose sentiment scores fall below a defined threshold (typically below zero). This targeted filtering enables the analysis to focus on emotionally charged expressions of dissatisfaction, which are more likely to contain traces of brand hate. The filtered dataset is presented in an interactive table format, which includes the review text, assigned sentiment score, and labeled reason(s) for the complaint. This step ensures that subsequent emotion classification and hate detection stages are grounded in relevant and affectively negative content, thus improving the precision of downstream analyses.

Figure 4 illustrates the sentiment-based filtering stage applied to the KFC illustration, where consumer reviews are automatically classified according to their emotional polarity. Reviews with negative sentiment scores are retained (370) and displayed in a structured table, along with associated complaint reasons and sentiment values. This output demonstrates how the artifact isolates affectively negative content, ensuring that subsequent emotion analysis and brand hate classification are grounded in relevant and emotionally charged consumer expressions.

4.2.2 Tokenization and Stopword Removal

In this step, each negatively valanced review is further preprocessed by removing common English stopwords (eg, "the", "and", "was", and "but") which typically carry little semantic weight. This tokenization and filtering process is performed using the tidytext and stopwords packages in R (Wickham and Bryan 2023). This step breaks each review into individual lowercase tokens (i.e., words), and a predefined stopword list is used to remove frequently occurring but semantically neutral terms. The result is a stopword-free version of each review, which retains only content words that contribute to emotional, evaluative, or thematic meaning. This refined version enhances the accuracy of subsequent emotion detection, hate classification, and keyword analysis by reducing noise and improving signal clarity. The Shiny interface presents both the cleaned (tokenized) and original review text side by side, allowing users to trace how preprocessing alters the structure and salience of emotional expressions (Fig. 5).

Fig. 4 Sentiment-based filtering of negative KFC reviews using the sentimentr package

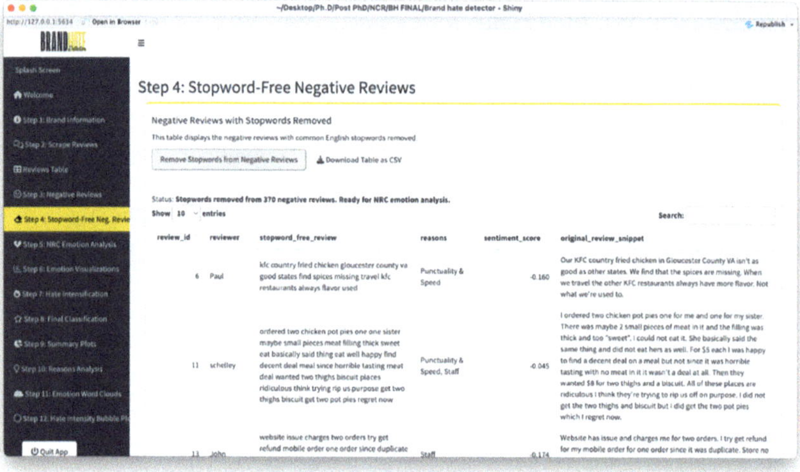

Fig. 5 Stopword removal and tokenization of negative KFC reviews

Once the data has been meticulously filtered and tokenized, we proceed to the analytical core of our pipeline. At this stage, advanced natural language processing techniques are employed to uncover the emotional landscape of the reviews and to classify the intensity and nature of brand hate expressed by consumers.

4.3 *Data Analysis*

The third stage centers on data analysis, beginning with the identification of five core negative emotions—anger, sadness, fear, disgust, and surprise—using the NRC Emotion Lexicon to quantify the emotional content of filtered reviews. This emotional profile is visualized through bar plots and used in conjunction with sentiment scores to classify brand hate intensity into mild, moderate, strong, or hybrid categories. By combining lexicon-based emotion detection with rule-based reclassification, this stage provides a multidimensional and scalable understanding of brand hate intensity in consumer feedback.

4.3.1 NRC Emotion Analysis of Negative Reviews

This stage of the pipeline identifies the emotional undercurrents embedded in the filtered negative reviews using the NRC Word-Emotion Association Lexicon (Mohammad and Turney 2013), with a targeted focus on five key negative emotions: anger, sadness, fear, surprise, and disgust. These categories align with the emotional frameworks proposed by Sternberg and Sternberg (2008), Kucuk (2019a), Zhang and Laroche (2020), which emphasizes that brand hate is not monolithic but instead driven by multiple affective experiences. Using the tidytext package, each stopword-free negative review is decomposed into individual tokens, which are matched against the NRC lexicon to count the occurrence of words associated with each emotion. The resulting emotional scores are tabulated for each review and displayed alongside the original review text and sentiment score. These quantified emotional components offer a nuanced representation of consumer negativity (Fig. 6).

4.3.2 Visualization of Emotion Distribution

To facilitate interpretation and provide a high-level summary of emotional content, the application generates a bar plot visualizing the total word count associated with each of the five detected emotions: sadness, fear, anger, disgust, and surprise. The data underlying the plot is derived from the NRC Emotion Lexicon analysis (Step 3.1), which tabulated the frequency of emotion-laden words in stopword-free negative reviews. This visualization, via ggplot2 package (Wickham and Bryan 2023), provides an immediate overview of the dominant emotional tone across the dataset (Fig. 7).

4.3.3 Brand Hate Intensification (Hybrid Classification Approach)

In this step, reviews are classified into three levels of brand hate—Mild, Moderate, and Strong—based on a hybrid classification method that combines qualitative

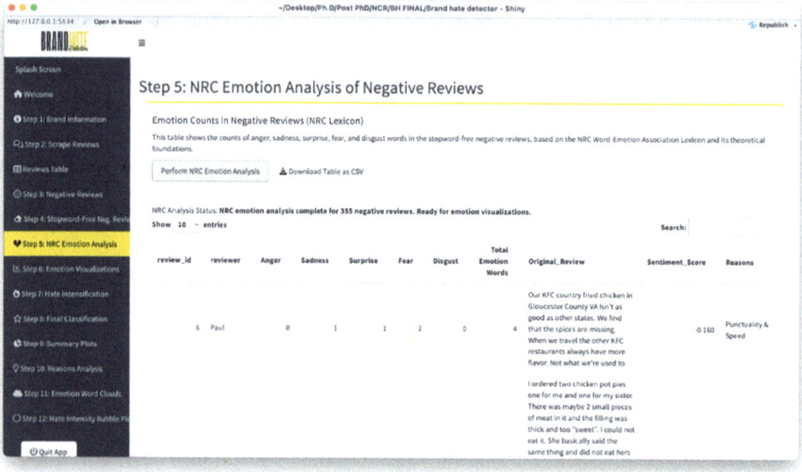

Fig. 6 NRC emotion analysis of negative reviews

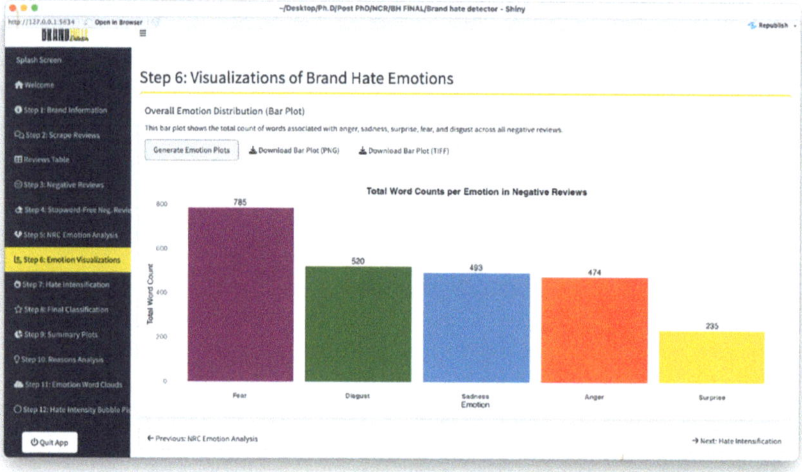

Fig. 7 Bar charts of brand hate primary emotions

emotion detection with quantitative sentiment analysis. This hybrid approach acknowledges that brand hate is a multidimensional construct rooted in distinct primary emotions and varying levels of intensity (Assoud and Berbou 2025a; Fetscherin 2019; Kucuk 2019a; Sternberg and Sternberg 2008; Zhang and Laroche 2020). Using the previous output from the NRC emotion lexicon, reviews are examined for the presence of key emotions: anger, sadness, surprise, fear, and

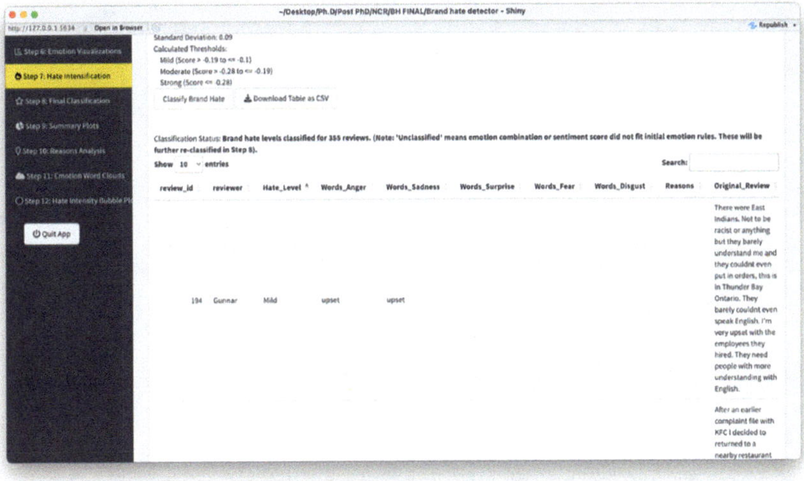

Fig. 8 Brand hate initial classification

disgust. Simultaneously, each review's sentiment score, computed using the sentimentr package, is evaluated relative to the dataset mean and standard deviation. This hybrid design mitigates the limitations of using sentiment score alone by incorporating emotional richness from the NRC lexicon. It also corrects for potential "paradoxical" cases—such as neutral sentiment scores with high emotional activation or vice versa—ensuring that the intensity of brand hate is captured holistically (Rinker 2017). By doing so, this step supports a more granular and scalable categorization of hostile consumer expressions, aligning with both psychological theories of hate and computational models of affective text analysis (Fig. 8).

4.3.4 Final Brand Hate Classification (Rule-Based Reclassification)

To further enhance the granularity of hate detection, reviews that did not conform to the standard thresholds in the previous step were re-evaluated through a rule-based procedure focused exclusively on emotional word co-occurrence (Mohammad and Turney 2013). This step served to reclassify ambiguous cases, now labeled as Hybrid Hate, rather than discarding them as unclassified. A review was categorized as Hybrid Hate if it exhibited emotional cues that partially overlapped with but did not fully satisfy the conditions for strong, moderate, or mild hate levels. For example, a review containing only one strong negative emotion (e.g., anger or disgust) without meeting multi-emotion thresholds could still reflect a potent yet idiosyncratic expression of brand hate. This refined labeling approach aligns with the multidimensional nature of hate theorized by Sternberg and Sternberg (2008) and acknowledges the hybrid forms of emotional hostility that consumers may exhibit in textual feedback. It also ensures

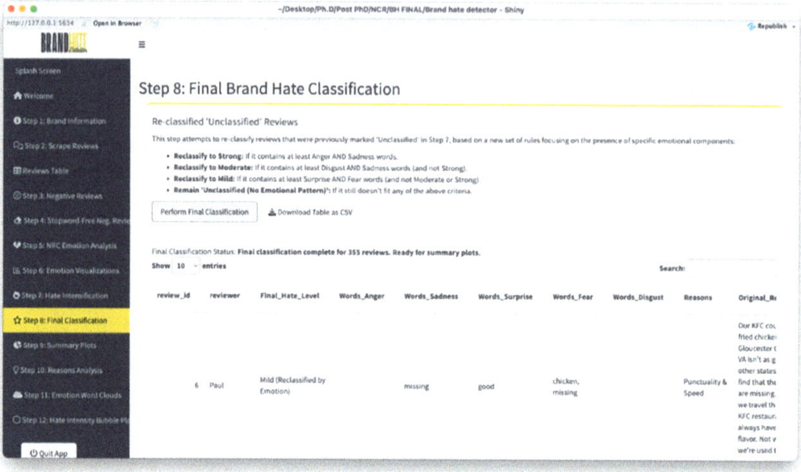

Fig. 9 Brand hate final classification

complete coverage of the dataset, preserving the emotional richness of reviews that might otherwise fall outside rigid classification rules (Fig. 9).

Having established a detailed emotional and categorical profile of brand hate within the dataset, the final stage focuses on synthesizing and communicating these insights. Through a suite of visualizations and summary reports. We translate complex analytical outcomes into actionable knowledge for both academic and managerial audiences.

4.4 Data Visualization and Reporting

The fourth stage focuses on data visualization and reporting to enhance interpretability and support evidence-based insights. It synthesizes brand hate classification results through bar charts, pie charts, word clouds, and a sentiment-based bubble plot—offering both quantitative summaries and qualitative emotional patterns across hate intensity levels. These visual outputs enable a comprehensive understanding of brand hate prevalence, underlying reasons, and emotional language, benefiting both academic inquiry and managerial decision-making.

4.4.1 Brand Hate Summary Visualizations

The ninth step of the analytical pipeline focuses on synthesizing the results of the brand hate classification process through clear visual representations. This stage

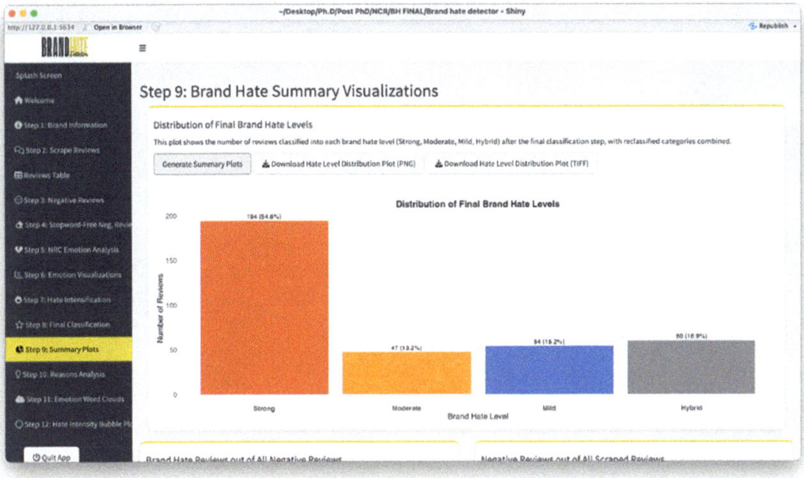

Fig. 10 Brand hate summary visualizations

provides aggregate insights into the distribution of reviews across the four brand hate intensity categories: Strong, Moderate, Mild, and Hybrid Hate. A bar chart summarizes the number and proportion of reviews in each category, while two pie charts present the overall proportion of brand hate reviews among all negative reviews (via ggplot2 and scales packages), as well as the share of negative reviews within the full scraped dataset. These summary visualizations not only verify the internal consistency of the classification logic but also offer a high-level understanding of the prevalence and variation of brand hate intensity in consumer feedback. The visual outputs enhance interpretability for both academic research and managerial decision-making (Fig. 10).

4.4.2 Reasons per Brand Hate Type Analysis

This step focuses on categorizing the primary justifications provided by consumers for their brand hate expressions across different intensity levels (Strong, Moderate, Mild, Hybrid). Each review contains a pre-tagged 'reason' field, which is aggregated and analyzed using frequency-based visualizations. The analysis is segmented by hate level, allowing the identification of reason patterns specific to each emotional intensity. This structured breakdown supports the methodological objective of linking emotional responses to concrete experiential triggers. The frequency counts are displayed through bar plots, providing a comparative view of the most cited reasons per category. This step enables a more nuanced understanding of the contextual drivers behind brand hate, aligning the classification with operational or service-related shortcomings from the consumer perspective (Fig. 11).

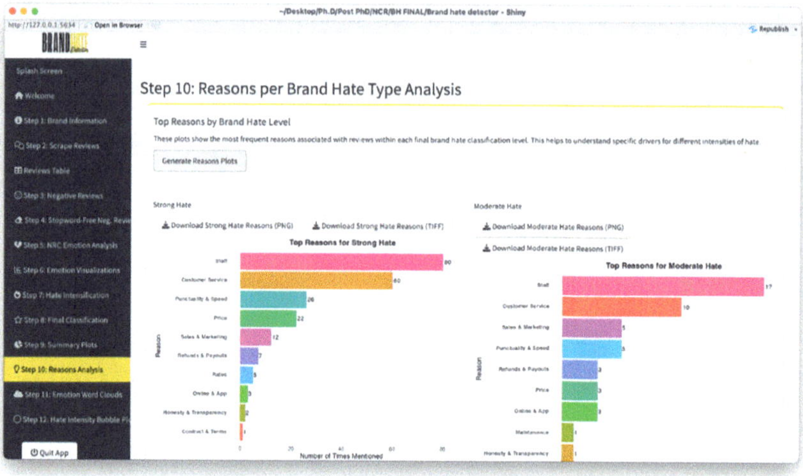

Fig. 11 Reasons per brand hate type analysis

4.4.3 Emotion Word Clouds by Brand Hate Type

To gain a visual understanding of the dominant emotional lexicon across brand hate categories, Step 11 employs word cloud visualizations. Word clouds provide an intuitive and accessible method to highlight the frequency of emotionally charged words found within consumer reviews previously categorized as Strong, Moderate, Mild, or Hybrid Hate in Step 8. Using the wordcloud package in R (Fellows et al. 2018), the analysis displays high-frequency terms in larger font sizes, allowing for immediate recognition of salient emotional expressions. Additionally, color palettes from RColorBrewer package (Wickham and Bryan 2023) were used to distinguish word clusters and enhance interpretability. Each word cloud is generated from the filtered set of reviews per hate category and reflects the specific lexical choices consumers use to articulate brand hate emotions, offering a qualitative complement to earlier sentiment classification steps. This visualization aids in identifying semantic patterns unique to each level of brand hate intensity and supports the triangulation of findings derived from quantitative text mining procedures (Fig. 12).

4.4.4 Hate Intensity Bubble Plot (Sentiment Versus Type)

In the final step of the analysis pipeline, a bubble plot was generated to visualize the distribution of reviews according to their brand hate classification and corresponding sentiment score. Each bubble represents a single review, positioned along the x-axis by its classified hate level (step 8) and along the y-axis by its sentiment score, as computed by the sentimentr package (step 3). To ensure clarity in visual

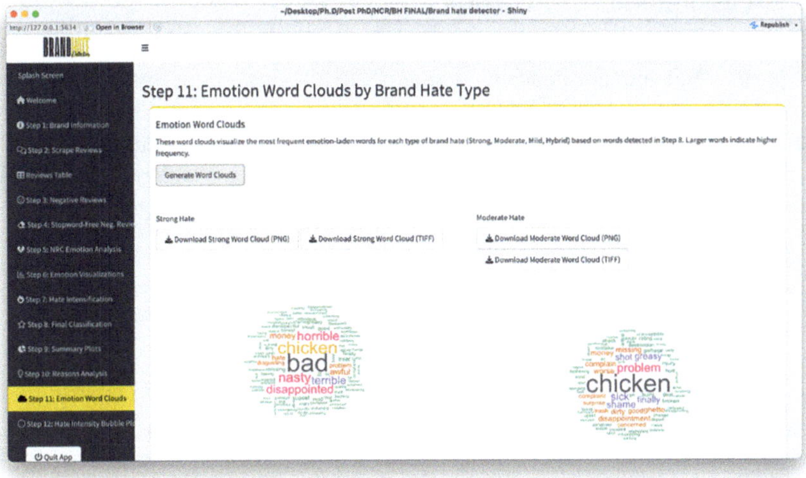

Fig. 12 Emotion word clouds by brand hate type

representation, jittering was applied to minimize overlap between data points. This visualization method provides a dual-layer view of the data, combining categorical hate classifications with continuous sentiment intensity. It serves as an integrative tool to validate and explore the internal consistency between qualitative emotion tagging (based on NRC lexicon analysis) and the quantitative polarity of reviewer sentiment (Fig. 13).

Fig. 13 Hate intensity bubble plot

5 Discussion and Conclusion

This paper offers a major conceptual and methodological contribution to the study of brand hate, relying on a hybrid pipeline that combines emotional lexicon analysis with sentiment analysis. This approach enables the operationalization of the multi-dimensionality of brand hate, enriches the understanding of underlying emotional dynamics, and opens new perspectives for both research and managerial practice.

5.1 Theoretical Contributions

One of the central theoretical contributions of this pipeline lies in its ability to validate and illustrate the complexity of brand hate, as conceptualized par Sternberg and Sternberg (2008), Zhang and Laroche (2020), Kucuk (2019a), and others. By distinguishing between different intensities (moderate, strong, hybrid) and forms of hate, the model links detected emotions to robust theoretical typologies. This granularity provides a nuanced interpretation of rejection phenomena, highlighting the diversity of emotional justifications (anger, disgust, sadness, fear, surprise) and their articulation with existing models. Thus, the pipeline moves beyond a monolithic view of brand hate, revealing the richness of affective dynamics and underlying logic.

5.2 Methodological Contributions

From a methodological standpoint, the principal innovation resides in the combination of a validated emotional lexicon with sentiment analysis tools, offering a transparent and interpretable alternative to "black box" models such as GPT-2. This hybridization ensures traceability of results, explains model decisions, and facilitates their adoption by both researchers and practitioners. The robustness of the system is reinforced by cross-validation with multiple artificial intelligence systems, ensuring the reproducibility and reliability of classifications. This methodological framework lays the groundwork for standardizing brand hate analyses, while remaining open to the integration of new tools, lexicons, and data sources.

5.3 Managerial Implications

The implications for brand management are significant. The developed pipeline enables managers to benefit from differentiated emotional diagnostics, providing early alerts on hate dynamics even in the absence of specific case studies. This ability to anticipate and qualify brand hate constitutes a strategic lever for adapting responses, targeting interventions, and preventing the escalation of reputational crises. The tool supports proactive management, based on a refined understanding of the emotions and justifications expressed by consumers, and paves the way for increased personalization of customer relationship policies.

5.4 Limitations and Future Research

Despite its contributions, the proposed detection tool is subject to several methodological limitations that should be acknowledged. First, the analysis is restricted to English-language reviews, which limits the international scope of the model and constrains its ability to capture culturally embedded emotional expressions and linguistic nuances. Second, the exclusive reliance on ConsumerAffairs as a single data source restricts the representativeness of the findings and limits their generalizability to other digital environments characterized by different interaction norms, platform architectures, and forms of consumer expression. In addition, the current design does not incorporate a temporal dimension, preventing the analysis of how brand hate evolves over time or how escalation and resolution dynamics unfold (Assoud and Berbou 2025a; Zarantonello et al. 2018). The pipeline also focuses primarily on brand hate stemming from direct negative consumption experiences, thereby overlooking ideological, symbolic, or collectively constructed forms of rejection that may be equally influential in certain sectors (Assoud and Berbou 2025a). Finally, empirical validation remains limited, as the artifact has not yet been applied to multiple concrete cases or confronted with data from varied industries and cultural contexts, which constrains the scope of the conclusions that can be drawn.

These limitations open several promising directions for future research and development. Extending the pipeline to multilingual contexts would broaden its applicability and allow cross-cultural comparisons of brand hate expressions. Integrating data from diverse sources—such as social media platforms, online forums, and alternative review sites—would improve representativeness and capture a wider range of discursive environments. Future work should also incorporate longitudinal mechanisms capable of tracking the evolution of brand hate over time, identifying emotional cycles and critical tipping points. Expanding the framework to include ideological, symbolic, and collective forms of hate would contribute to a more comprehensive understanding of rejection phenomena beyond direct consumption experiences. Moreover, the development of predictive and experimental extensions could enable the anticipation of reputational crises, comparative analyses of emotional discourse across competing brands, and the evaluation of specific managerial response strategies. Applying the model across varied sectors and cultural contexts would further test its robustness and refine its theoretical and practical contributions.

In summary, this work lays the foundations for an integrated, transparent, and explainable approach to brand hate detection. While acknowledging its current limitations, it opens multiple avenues for future research and managerial innovation, and invites critical reflection on the conditions under which emotional analysis tools can be generalized, adapted, and meaningfully appropriated within the field of brand management.

References

Abbasi, A.Z., Fayyaz, M.S., Ting, D.H., Munir, M., Bashir, S., Zhang, C.: The moderating role of complaint handling on brand hate in the cancel culture. Asia Pac. J. Bus. Adm. **15**(1), 46–71 (2023). https://doi.org/10.1108/APJBA-06-2021-0246

Albladi, A., Islam, M., Seals, C.: Sentiment analysis of Twitter data using NLP models: a comprehensive review. IEEE Access **13**, 30444–30468 (2025). https://doi.org/10.1109/access.2025.3541494

Arvidsson, A.: Brands: Meaning and Value in Media Culture. Routledge (2006)

Assoud, M., Berbou, L.: Conceptualizing brand hate escalation: a dual SEM-PLS/necessary condition analysis approach. Afr. J. Hum. Resour. Mark. Organ. Stud. **2**(1), 55–81. https://hdl.handle.net/10520/ejc-aa_ajhrmos_v2_n1_a4 (2025a)

Assoud, M., Berbou, L.: Mapping the evolution of brand hate: a comprehensive bibliometric analysis. J. Consum. Satisf. Dissatisfaction Complain. Behav. **38**(1), 21–58. https://www.jcsdcb.com/index.php/JCSDCB/article/view/1086 (2025b)

Assoud, M., Berbou, L.: The dynamics of brand hate during a social-political conflict: empirical evidence from Moroccan consumers during the 2023 war on Gaza. In: Marcao, R., Santos, V., Pinhal, R. (eds.) Reinventing Corporate Strategy for Competitive Advantage, pp. 169–204. IGI Global Scientific Publishing (2026a). https://doi.org/10.4018/979-8-3373-5428-6.ch007

Assoud, M., Berbou, L.: The impact of political brand hate on the re-vote intention: a moderating role of performance tolerance. In: Panagiotidou, G. (ed.) Understanding Voter Behavior with Predictive Modeling (pp. 359–398). IGI Global Scientific Publishing (2026b). https://doi.org/10.4018/979-8-3373-1037-4.ch012

Assoud, M., Berbou, L., Vieira, L.S.: Brand hate semiotics: hate as a story theory. Netnographic approach during the war on Gaza. J. Islam. Mark., 1–35 (2025a). https://doi.org/10.1108/jima-11-2023-0382

Assoud, M., Berbou, L., Vieira, L.S.: Towards a theory of brand hate: empirical evidence of mediterranean countries. J. Consum. Satisf. Dissatisfaction Complain. Behav. **38**(2), 85–114. https://www.jcsdcb.com/index.php/JCSDCB/article/view/1274 (2025b)

Barlow, M.A., Verhaal, J.C., Angus, R.W.: Optimal distinctiveness, strategic categorization, and product market entry on the Google Play app platform. Strateg. Manag. J. **40**(8), 1219–1242 (2019). https://doi.org/10.1002/smj.3019

Bayarassou, O., Becheur, I., Valette-Florence, P.: "Fight or flight": coping responses to brand hate. J. Prod. Brand. Manag. **30**(3), 492–511 (2020). https://doi.org/10.1108/JPBM-08-2019-2519

Brandao, A., Popoli, P.: "I'm hatin' it"! Negative consumer-brand relationships in online anti-brand communities. Eur. J. Mark. **56**(2), 622–650 (2022). https://doi.org/10.1108/ejm-03-2020-0214

Brandão, A., Ribeiro, B., Gadekar, M.: You ruined our love story, but I just cannot hate you—a moderation–mediation analysis of past experienced brand love and brand hate. Int. J. Consum. Stud. **47**(3), 1126–1138 (2023). https://doi.org/10.1111/ijcs.12892

Bryson, D., Atwal, G., Hultén, P.: Towards the conceptualisation of the antecedents of extreme negative affect towards luxury brands. Qual. Mark. Res. **16**(4), 393–405 (2013). https://doi.org/10.1108/QMR-06-2013-0043

Calomiris, C.W., Mamaysky, H.: How news and its context drive risk and returns around the world. J. Financ. Econ. **133**(2), 299–336 (2019). https://doi.org/10.1016/j.jfineco.2018.11.009

Carroll, B.A., Ahuvia, A.C.: Some antecedents and outcomes of brand love. Mark. Lett. **17**, 79–89 (2006). https://doi.org/10.1007/s11002-006-4219-2

Costa, C., Azevedo, A.: Antecedents and consequences of the "cancel culture" firestorm journey for brands: is there a possibility for forgiveness? J. Mark. Manag. **40**(3–4), 289–312 (2024). https://doi.org/10.1080/0267257X.2023.2266465

Costa, J., Azevedo, A.: Beyond brand hate: are you willing to forgive a negative experience with your smartphone mobile communications' brand? Corp. Reput. Rev. **26**(2), 83–96 (2023). https://doi.org/10.1057/s41299-022-00139-7

Curina, I., Francioni, B., Hegner, S.M., Cioppi, M.: Brand hate and non-repurchase intention: a service context perspective in a cross-channel setting. J. Retail. Consum. Serv. **54**, 102031 (2020). https://doi.org/10.1016/j.jretconser.2019.102031

Curtis, T., Abratt, R., Rhoades, D., Dion, P.: Customer loyalty, repurchase and satisfaction: a meta-analytical review. J. Consum. Satisf. Dissatisfaction Complain. Behav. 1–26, https://jcsdcb.com/index.php/JCSDCB/article/view/14 (2011)

Dege, D., Brüggemann, P.: Marketing analytics with RStudio: a software review. J. Mark. Anal. **12**(2), 465–470 (2024). https://doi.org/10.1057/s41270-023-00264-0

Doi, J., Potter, G., Wong, J., Alcaraz, I., Chi, P.: Web application teaching tools for statistics using R and shiny. Technol. Innov. Stat. Educ. **9**(1), 1–33 (2016). https://doi.org/10.5070/t591027492

Dyer, T.A., Lang, M.H., Stice-Lawrence, L.: Discussion of: 10-K disclosure repetition and managerial reporting incentives. J. Financ. Report. **2**(1), 133–138 (2017). https://doi.org/10.2308/jfir-51950

Farhat, Z., Chaney, D.: Introducing destination brand hate: an exploratory study. Curr. Issue Tour. **24**(17), 2472–2488 (2021). https://doi.org/10.1080/13683500.2020.1844160

Fellows, I., Fellows, M.I., Rcpp, L., Rcpp, L.: Package 'wordcloud.' R Packag. Version **2**, 331 (2018)

Fetscherin, M.: The five types of brand hate: how they affect consumer behavior. J. Bus. Res. **101**, 116–127 (2019). https://doi.org/10.1016/j.jbusres.2019.04.017

Feuerriegel, S., Maarouf, A., Bär, D., Geissler, D., Schweisthal, J., Pröllochs, N., Robertson, C.E., Rathje, S., Hartmann, J., Mohammad, S.M., Netzer, O., Siegel, A.A., Plank, B., Van Bavel, J.J.: Using natural language processing to analyse text data in behavioural science. Nat. Rev. Psychol. **4**(2), 96–111 (2025). https://doi.org/10.1038/s44159-024-00392-z

Fournier, S.: Consumers and their brands: developing relationship theory in consumer research. J. Consum. Res. **24**(4), 343–373 (1998). https://doi.org/10.1086/209515

Gois, A.M., Moura, L.R.C., Goncalves, C., Chinelato, F.B.: Hate towards brands of educational institutions: she hates for ideology, and he hates for previous experiences. J. Appl. Res. High. Educ. **21**, 0158 (2022). https://doi.org/10.1108/jarhe-04-m21-0158

Hashim, S., Kasana, S.: Antecedents of brand hate in the fast food industry. Span. J. Mark. ESIC **23**(2), 227–248 (2019). https://doi.org/10.1108/SJME-10-2018-0047

Hegner, S.M., Fetscherin, M., van Delzen, M.: Determinants and outcomes of brand hate. J. Prod. Brand. Manag. **26**(1), 13–25 (2017). https://doi.org/10.1108/JPBM-01-2016-1070

Hirschman, A.O.: Exit, Voice, and Loyalty: Responses to Decline in Firms, Organizations, and States, Vol. 25. Harvard University Press (1970)

Husnain, M., Syed, F., Akhtar, W., Usman, M.: Effects of brand hate on brand equity: the role of corporate social irresponsibility and similar competitor offer. Mark. Manag. Innov. **3**, 75–86 (2020). https://doi.org/10.21272/mmi.2020.3-06

Islam, T., Attiq, S., Hameed, Z., Khokhar, M.N., Sheikh, Z.: The impact of self-congruity (symbolic and functional) on the brand hate: a study based on self-congruity theory. Br. Food J. **121**(1), 71–88 (2019). https://doi.org/10.1108/BFJ-03-2018-0206

Ito, T.A., Larsen, J.T., Smith, N.K., Cacioppo, J.T.: Negative information weighs more heavily on the brain: the negativity bias in evaluative categorizations. J. Pers. Soc. Psychol. **75**(4), 887 (1998)

Kang, Y., Cai, Z., Tan, C.-W., Huang, Q., Liu, H.: Natural language processing (NLP) in management research: a literature review. J. Manag. Anal. **7**(2), 139–172 (2020). https://doi.org/10.1080/23270012.2020.1756939

Kucuk, S.U.: Brand Hate: Navigating Consumer Negativity in the Digital World. Springer International Publishing (2019a). https://doi.org/10.1007/978-3-030-00380-7
Kucuk, S.U.: Consumer brand hate: steam rolling whatever I see. Psychol. Mark. **36**(5), 431–443 (2019b). https://doi.org/10.1002/mar.21175
Kucuk, S.U.: Consumer Voice: The Democratization of Consumption Markets in the Digital Age. Springer International Publishing (2020). https://doi.org/10.1007/978-3-030-53983-2
Lis, B., Fischer, M.: Analyzing different types of negative online consumer reviews. J. Prod. Brand. Manag. **29**(5), 637–653 (2020). https://doi.org/10.1108/jpbm-05-2018-1876
Liu, H., Wu, S.T., Li, D., Jonnalagadda, S., Sohn, S., Wagholikar, K., Haug, P.J., Huff, S.M., Chute, C.G.: Towards a semantic lexicon for clinical natural language processing. In: AMIA Annual Symposium Proceedings (2012)
Loureiro, S.M.C., Kaufmann, H.R.: The role of online brand community engagement on positive or negative self-expression word-of-mouth. Cogent Bus. Manag. **5**(1), 1508543 (2018). https://doi.org/10.1080/23311975.2018.1508543
Manning, C., Schutze, H.: Foundations of Statistical Natural Language Processing. MIT Press (1999)
McDonald, D.D.: Natural language generation. Handb. Nat. Lang. Process. **2**, 121–144 (2010)
Mednini, L., Noubigh, Z., Turki, M.D.: Natural language processing for detecting brand hate speech. J. Telecommun. Digit. Econ. **12**(1), 486–509 (2024). https://doi.org/10.18080/jtde.v12n1.859
Melumad, S., Inman, J.J., Pham, M.T.: Selectively emotional: how smartphone use changes user-generated content. J. Mark. Res. **56**(2), 259–275 (2019). https://doi.org/10.1177/0022243718815429
Mohammad, S.M., Turney, P.D.: Crowdsourcing a word–emotion association Lexicon. Comput. Intell. **29**(3), 436–465 (2013). https://doi.org/10.1111/j.1467-8640.2012.00460.x
Odoom, R., Kosiba, J.P.B., Odoom, P.T.: Brand hate experiences and the role of social media influencers in altering consumer emotions. J. Brand Manag. **14**, 529–542 (2024). https://doi.org/10.1057/s41262-024-00356-x
Oliver, R.L.: A cognitive model of the antecedents and consequences of satisfaction decisions. J. Mark. Res. **17**(4), 460–469 (1980). https://doi.org/10.1177/002224378001700405
Pan, Y., Huang, P., Gopal, A.: Storm clouds on the horizon? New entry threats and R&D investments in the U.S. IT industry. Inf. Syst. Res. **30**(2), 540–562 (2019). https://doi.org/10.1287/isre.2018.0816
Parveen, U.: An analysis of linguistics patterns in online product or service reviews and their influence on customer behavior. Qlantic J. Soc. Sci. Hum. **6**(2), 111–117 (2025). https://doi.org/10.55737/qjssh.vi-ii.25358
Pinto, O., Brandão, A.: Antecedents and consequences of brand hate: empirical evidence from the telecommunication industry. Eur. J. Manag. Bus. Econ. **30**(1), 18–35 (2021). https://doi.org/10.1108/EJMBE-04-2020-0084
Rinker, T.: Package 'sentimentr.' Retrieved **8**, 31 (2017)
Romani, S., Grappi, S., Dalli, D.: Emotions that drive consumers away from brands: measuring negative emotions toward brands and their behavioral effects. Int. J. Res. Mark. **29**(1), 55–67 (2012). https://doi.org/10.1016/j.ijresmar.2011.07.001
Sajid, M., Zakkariya, K.A., Suki, N.M., Islam, J.U.: When going green goes wrong: the effects of greenwashing on brand avoidance and negative word-of-mouth. J. Retail. Consum. Serv. **78**, 103773 (2024). https://doi.org/10.1016/j.jretconser.2024.103773
Sameeni, M.S., Qadeer, F., Ahmad, W., Filieri, R.: An empirical examination of brand hate influence on negative consumer behaviors through NeWOM intensity. Does consumer personality matter? J. Bus. Res. **173**, 114469 (2024). https://doi.org/10.1016/j.jbusres.2023.114469
Santos, C., Coelho, A., Marques, A.: The greenwashing effects on corporate reputation and brand hate, through environmental performance and green perceived risk. Asia Pac. J. Bus. Adm. **16**(3), 655–676 (2024). https://doi.org/10.1108/APJBA-05-2022-0216
Sharma, I., Jain, K., Behl, A.: Motives of the self and brand hate. J. Consum. Mark. **39**(7), 708–725 (2022). https://doi.org/10.1108/jcm-04-2021-4635

Shaver, P., Schwartz, J., Kirson, D., O'connor, C.: Emotion knowledge: further exploration of a prototype approach. J. Pers. Soc. Psychol. **52**(6), 1061 (1987)

Sternberg, R.J., Sternberg, K.: The Nature of Hate. Cambridge University Press (2008)

Syauqi, A.L., Wibawa, A.P.: Advances in machine transliteration methods, limitations, challenges, applications and future directions. Nat. Lang. Process. J. **11**, 100158 (2025). https://doi.org/10.1016/j.nlp.2025.100158

Valenzuela, C., Castellucci, L., Mena, M.T., Bianchi, C.: Consumer brand hate: the role of ambivalence. Int. Rev. Retail. Distrib. Consum. Res. **32**(1), 100–125 (2022). https://doi.org/10.1080/09593969.2021.2018017

Wickham, H., Bryan, J.: R Packages. O'Reilly Media, Inc. (2023)

Xin Ying Ong, S., Vila-Lopez, N.: Brand hate in the cosmetic industry: drugstore versus luxury make-up brands. Young Consum. **25**(1), 109–127 (2024). https://doi.org/10.1108/YC-10-2022-1624

Yadav, A., Chakrabarti, S.: Brand hate: a systematic literature review and future research agenda. Int. J. Consum. Stud. **46**(5), 1992–2019 (2022). https://doi.org/10.1111/ijcs.12772

Yang, J., Mundel, J.: Effects of brand feedback to negative eWOM on brand love/hate: an expectancy violation approach. J. Prod. Brand. Manag. **31**(2), 279–292 (2022). https://doi.org/10.1108/JPBM-05-2020-2900

Yildirim, G., Kübler, R.: Applied Marketing Analytics Using R. SAGE Publications Limited (2023)

Yu, J.H., Chauhan, D.: Trends in NLP for personalized learning: LDA and sentiment analysis insights. Educ. Inf. Technol. **30**(4), 4307–4348 (2025). https://doi.org/10.1007/s10639-024-12988-2

Zarantonello, L., Romani, S., Grappi, S., Bagozzi, R.P.: Brand hate. J. Prod. Brand. Manag. **25**(1), 11–25 (2016). https://doi.org/10.1108/jpbm-01-2015-0799

Zarantonello, L., Romani, S., Grappi, S., Fetscherin, M.: Trajectories of brand hate. J. Brand Manag. **25**(6), 549–560 (2018). https://doi.org/10.1057/s41262-018-0105-5

Zhang, C., Laroche, M.: Brand hate: a multidimensional construct. J. Prod. Brand. Manag. **30**(3), 392–414 (2020). https://doi.org/10.1108/JPBM-11-2018-2103

Zhang, Q., Su, L., Zhou, L., Dai, Y.: Distance brings about beauty: when does the influence of positive travel online reviews grow stronger relative to negative reviews? J. Travel Res. **64**(1), 172–188 (2025). https://doi.org/10.1177/00472875231209426

Digitalization of Financial and Accounting Processes in Public Institutions: A Case Study of a Public Organization in Casablanca, Morocco

Abdelmounim Bouziane

Abstract The digitalization of accounting and financial processes, as well as the implementation of Enterprise Resource Planning (ERP) systems, represent major trends in organizations worldwide. However, public institutions often face challenges in adopting such integrated systems due to structural, procedural, and resource constraints. This study focuses on a public institution in Casablanca (Morocco) that continues to manage its financial and accounting operations using traditional tools such as Microsoft Excel. Through a qualitative methodology combining direct observations, interviews with managers, and questionnaires administered to all staff, the research identifies the limitations, inefficiencies, and risks associated with manual processes, including errors, data fragmentation, and delays in financial decision-making. The study further explores the factors influencing the potential success of ERP integration, highlighting the importance of IT support, staff digital competencies, interdepartmental collaboration, and the management of change. Based on these findings, a practical, step-by-step action plan is proposed to implement a suitable ERP system and update financial and accounting procedures. This plan aims to enhance operational efficiency, ensure data reliability, and strengthen financial governance within public institutions.

Keywords Digitalization · ERP systems · Public sector

1 Introduction

In recent years, the digitalization of financial and accounting processes has become a central theme in organizational transformation, reshaping the way financial information is collected, processed, and reported. The introduction of Enterprise Resource Planning (ERP) systems has enabled organizations to integrate various business

A. Bouziane (✉)
Chouaib Doukkali University & Urban Agency of Casablanca, El Jadida, Morocco
e-mail: bouziane.a@ucd.ac.ma

A. Bouziane and K. Bouziane (eds.), *AI, Management, and Culture: Shaping Organizations in the Digital Age*, Studies in Computational Intelligence 1291,
https://doi.org/10.1007/978-3-032-27882-1_3

functions—such as accounting, finance, procurement, and human resources—into a unified digital platform. This integration has improved the efficiency, reliability, and transparency of financial operations, particularly in private sector organizations. However, in the public sector, the adoption of ERP systems and digital financial management tools has been comparatively slower. Many public institutions continue to rely on traditional tools such as spreadsheets and manual data entry, which can lead to inefficiencies, data inconsistencies, and increased operational risks. The reluctance to adopt digital systems in public institutions can be attributed to several factors, including budgetary constraints, lack of technical expertise, resistance to change, and complex administrative procedures. This study focuses on a public institution in Casablanca, Morocco, that continues to manage its accounting and financial operations through conventional tools. By analyzing this case, the research aims to identify the shortcomings of the current system, assess the potential benefits of adopting digital financial solutions, and propose a practical framework for implementing an ERP system adapted to the institution's needs. The main objectives of this study are threefold:

- To evaluate the current state of digitalization within the institution's accounting and financial processes.
- To identify the operational and governance challenges associated with traditional financial management tools.
- To propose a digital transformation roadmap that enhances financial efficiency, reliability, and transparency in public institutions.

Ultimately, this study seeks to contribute to the growing body of knowledge on public sector digital transformation by providing practical insights into how Moroccan public institutions can transition toward integrated digital systems that foster better governance, transparency, and accountability. In this context, the research addresses the following central question.

To what extent can the integration of an ERP system into the financial and accounting management of public organizations enhance operational performance, strengthen internal control, and mitigate financial and organizational risks?

By exploring this question through a detailed case study of a public institution in Casablanca, the study aims to identify both the barriers and the success factors influencing the digital transformation of financial management within the public sector.

2 Literature Review

The digital transformation of accounting and financial management has today become an essential pillar of organizational modernization on a global scale. This digitalization is not limited to the introduction of new technologies; it also represents a profound managerial and cultural shift, capable of redefining internal processes, improving the quality of decision-making, and reshaping professional practices. In this context, Enterprise Resource Planning systems, commonly known as ERP, are positioned as key strategic levers. These systems make it possible to coherently integrate all financial and administrative functions, optimize operational efficiency, and ensure greater reliability of data.

In the private sector, the adoption of ERPs is closely correlated with a significant improvement in financial reporting, the strengthening of internal control, and real-time access to key performance indicators. ERP solutions facilitate the automation of repetitive accounting tasks, reduce data duplication, improve the accuracy of records, and ensure compliance with international accounting and tax standards. Moreover, they promote the traceability of operations and limit human intervention, thereby contributing to better governance and a reduction in the risk of error or fraud.

In organizations in developed countries, ERPs are designed to adapt to the specific structure of the company through configurable modules covering all organizational functions. In the field of accounting, research has mainly focused on management accounting, often leaving aside the application of legal standards (accounting or tax). Studies on auditing and financial accounting are also part of this context, favoring the analysis of practices within highly standardized and technologically advanced environments. With regard to management control, it has been shown that ERP, on its own, does not constitute a determining factor in the evolution of the management controller's role. In fact, pre-existing tools, models, and control procedures are often maintained, with ERPs mainly serving to optimize and refine these traditional practices rather than radically transforming them.

On the accounting level, ERPs offer multiple practical advantages. They allow real-time recording of transactions, with precise monthly breakdowns and the possibility of monitoring accounting balances at any time. ERP systems also facilitate the instant extraction of summary financial statements, automatically flag anomalies, and can even prevent certain fraudulent manipulations. Most importantly, they guarantee the traceability of operations, limit human intervention, and strengthen the reliability and transparency of financial information.

However, despite the numerous advantages offered by ERP systems, the public sector often remains behind in their effective implementation. This lag can be explained by the presence of multiple structural, organizational, legislative, and institutional obstacles that slow down the digital transformation of public administrations. Several research studies have highlighted the particular complexity of adopting ERPs in the public context, where centralized decision-making power, excessive process hierarchization, regulatory rigidity, and budgetary constraints strongly limit the capacity for innovation and adaptation.

Indeed, unlike the private sector, public organizations must contend with an administrative environment characterized by low flexibility and a control culture that prioritizes compliance over performance. The introduction of an ERP therefore implies a profound redefinition of management practices, decision-making channels, and internal responsibilities, which can generate resistance to change and slow down the appropriation process.

Furthermore, the successful integration of ERP systems in the public sector depends closely on a set of human, institutional, and organizational factors. Among these are user training and awareness, active support from public authorities and political decision-makers, the modernization of the administration, and the implementation of a comprehensive change management strategy.

Fernandez et al. (2018) conducted a study on the use of ERP systems in the public sector, with a particular focus on their implications for accounting and financial functions. The authors emphasize that the integration of an ERP in public administrations is not limited to a technological innovation, but constitutes a genuine organizational transformation project, directly affecting budget management, public accounting, and financial control.

According to their conclusions, the implementation of an ERP system aims primarily to improve the reliability, traceability, and transparency of accounting and financial information. However, this process remains confronted with several structural and organizational challenges. Leaders must anticipate the risks linked to system complexity, resistance to change, and coordination between the various accounting, financial, and administrative departments. The authors therefore stress the need for any public institution, prior to adoption, to carry out an in-depth analysis of its strengths and weaknesses in order to assess its readiness from technical, managerial, and financial perspectives.

Fernandez et al. (2018) also recommend that implementation programs be realistic and gradual, based on precise planning and a balance between modernization and continuity of accounting processes. An ERP that is poorly adapted or deployed without adequate anticipation can generate disruptions in the accounting processing chain and compromise the quality of financial reporting. Conversely, a well-structured implementation makes it possible to automate repetitive operations, strengthen internal control mechanisms, and ensure compliance with applicable accounting and budgetary standards.

Along the same lines, Fernandez et al. (2017) carried out an in-depth study on the use and impact of ERP systems in the public sector, more specifically within local authorities in Malaysia. Their research aims to better understand the effects of digitalization on the accounting and financial performance of public institutions.

The results of their study show that the adoption of an ERP system is an effective strategy for improving the reliability of accounting information, strengthening financial transparency, and increasing the operational efficiency of public organizations. By automating data entry, consolidation, and reporting processes, ERP promotes better integration between the various administrative and financial functions, while contributing to more rigorous budget management and more informed decision-making.

Fernandez et al. (2017) conclude that the implementation of an ERP in Malaysian local authorities has made it possible to rationalize information flows, strengthen internal control, and reduce human errors in accounting processes. These results encourage public institutions to adopt similar strategies in order to improve financial governance and the overall performance of the public sector.

In the Moroccan context, digitalization is currently expanding progressively across all sectors and fields of activity. This dynamic reflects a national will to modernize public services and improve organizational efficiency through the adoption of digital technologies.

However, the integration of ERP systems in Moroccan public institutions is still in its early stages. Very few academic studies have focused on this subject, which limits the understanding of the challenges, opportunities, and impacts associated with these projects. Most digitalization efforts in the Moroccan public sector remain mainly concentrated on e-government and public finance reform, with particular emphasis on improving administrative procedures, dematerializing services, and monitoring public expenditure.

In this context, Kamhi and My Abdelouhab (2025) carried out a study on the integration of ERP systems in the public sector. Several important conclusions emerge from their work: the integration of ERPs in the public sector is not yet widespread in Morocco, it is a project that public institutions have only recently begun to launch, and research on this topic remains limited.

Despite these limitations, ERPs are beginning to be perceived as a strategic tool for modernizing the management of public institutions. Their adoption mainly concerns key areas such as public procurement management, budget planning, accounting, and financial management. In these areas, ERPs make it possible to automate processes, enhance the reliability of accounting and financial information, ensure transaction traceability, and strengthen transparency and internal control.

The authors also highlight that the successful implementation of an ERP in the public sector strongly depends on several factors: support from top management, active involvement of project managers, training and support for users, and change management tailored to the specific constraints of the public sector. These elements are essential to ensure that ERP systems do not merely automate procedures, but actually contribute to improving the performance and efficiency of the accounting and financial functions of public institutions.

Building on these findings, this research aims to fill an important gap in the literature by exploring how the implementation of ERP systems can transform accounting and financial management practices within Moroccan public institutions. Based on a case study carried out in a public organization in Casablanca, the objective is to provide a contextualized understanding of the risks associated with the use of traditional tools, the shortcomings and inefficiencies generated by the absence of an ERP, and the potential benefits and gains associated with its integration. The study also examines the technical and organizational factors that influence the success of digital transformation initiatives in the public sector.

3 Methodology

The methodology adopted in this study is based on a mixed approach, combining qualitative and quantitative tools in order to ensure a thorough understanding of financial and accounting management mechanisms and to assess the conditions for the successful integration of an ERP system.

3.1 Direct Observation Phase

In a first stage, on-site observation was carried out within several organizational units involved in the financial and accounting management process.

The units observed are:

- The department in charge of the budget
- The department in charge of accounting and cash management
- The department in charge of public procurement
- The department in charge of authorization/commitment
- And finally, the unit responsible for payments.

The observation consisted in assisting employees in the performance of their daily tasks in order to understand the processes in place, the interactions between departments, as well as the difficulties encountered.

Each unit was observed for five days, for a total duration of approximately twenty-five days for all observations.

3.2 Semi-structured Interview Phase

Following this first stage, semi-structured interviews were conducted with the heads of each of the observed units (five people in total).

These interviews aimed to deepen certain findings from the observation and to gather managers' perceptions regarding the limitations of the current system and the prospects for integrating an ERP into financial and accounting management. The interview guide is presented below (Table 1).

3.3 Design and Distribution of the Questionnaire

The information collected through observation and interviews was subjected to a narrative analysis. This analysis made it possible to identify the main themes and relevant variables, drawn both from the scientific literature and from field realities, within an adjusted positivist perspective. On this basis, a structured questionnaire was developed and distributed to all staff members of the structures concerned, including heads of department, i.e. a total of 13 people. The response rate was 100%, which reflects strong staff commitment to the research process. The administered questionnaire focused on the following research variables (Table 2).

Table 1 Interview guide—integration of an ERP into financial and accounting management

Introduction to the interview Presentation of the purpose of the research Assurance of the confidentiality of responses Explanation of how the interview will proceed (estimated duration: 30–45 min) Participant's consent
Context and work organization Could you briefly describe the main missions of your department? How is work organized within your team (information flows, coordination with other departments)? What IT tools are currently used in day-to-day management?
Budgetary, accounting and financial processes How does the budget forecasting process take place in your unit? What are the key stages in accounting processing and cash management? What difficulties do you encounter in coordination between departments (budget, commitment/authorization, public procurement, payment)?
Information system and existing tools Do you have an integrated computerized system for financial management? If yes, what are its main strengths and limitations? If not, how do you assess the effectiveness of the current tools (spreadsheets, stand-alone software, manual entry, etc.)?
Perceptions and expectations regarding an ERP In your view, what would be the expected benefits of implementing an ERP? What concerns or resistance do you anticipate within your department (training, adaptation, workload, etc.)? Which factors, based on your experience, are essential to ensure the success of such an integration?
Human resources and change support Does the staff have sufficient skills to use an ERP? What types of training or support would be necessary in your opinion? How do you perceive management's commitment to digital transformation projects?
Conclusion Would you like to add any point or suggestion concerning the modernization of the management system? Thank the participant for their availability and contribution

3.4 *Data Analysis Phase*

The collected data were processed through two levels of analysis.

A descriptive analysis, aimed at providing an overview of current practices, levels of mastery of processes, and perceptions related to the use of IT tools.

A linear regression analysis, intended to identify explanatory and determining factors likely to promote the successful integration of an ERP system into the organization's financial and accounting management.

This combined approach made it possible to link a qualitative understanding of organizational functioning with a quantitative validation of hypotheses derived from the field, thereby strengthening the robustness and credibility of the research results.

Table 2 Research variables

Dimension/variable	Definition/description	Conceptual references
Perceived usefulness	Degree to which the staff member perceives that using the ERP will improve their job performance	Bouziane (2025a, b)
Perceived ease of use	Degree to which the staff member believes that using the ERP will be easy and will not require significant effort	Bouziane (2025a, b)
Attitude toward use	Overall positive or negative feeling toward using the ERP	Fishbein and Ajzen (1975), Davis (1989)
Intention to use	Willingness of the staff member to actually use the ERP system in the context of their work	Venkatesh and Davis (2000)
Professional constraints	Obstacles related to workload, staff availability, or administrative procedures	Field data
Resistance to change	Degree of conscious or unconscious opposition to digital transformation and the adoption of new practices	Oreg (2003), Piderit (2000)
Training difficulties	Level of perceived difficulty in understanding and learning to use the ERP	Venkatesh et al. (2012), Ragu-Nathan et al. (2008)
Perceived organizational support	Level of commitment and support from management for the adoption of the ERP system	Eisenberger et al. (1986), Venkatesh et al. (2003)
Digital competence	Level of mastery of IT and digital tools by staff members	Bouziane and Bouziane (2023)
Organizational conditions	Set of resources and infrastructures available to support ERP implementation	Venkatesh et al. (2003), DeLone and McLean (2003)
Previous ERP project experience	Past history of the ERP project, including perceptions related to the previous failure	Field data

(continued)

Table 2 (continued)

Dimension/variable	Definition/description	Conceptual references
Quality of the relationship between functional units and IT	Reflects the level of cooperation, communication, and trust between the IT department and the other departments (budget, accounting, treasury, etc.) involved in the ERP project	Field data
Budget constraints and procurement procedures	Set of financial and regulatory limitations linked to the public nature of the organization, which influence the quality of the ERP system acquired and its implementation	Field data

4 Results

4.1 Descriptive Analysis of Results

The descriptive analysis is based on data from observations, interviews, and the questionnaire administered to the departments concerned. It highlights several findings relating to work organization, tools used, and difficulties encountered in budgetary, accounting, and financial management.

4.1.1 Organization and Data Management

Each department has its own database, mainly consisting of Excel files containing budgetary, accounting, and financial information. These files are not centralized and do not communicate with each other. Data are transmitted either by email or through the physical exchange of removable media such as USB sticks or CD-ROMs.

This sharing method creates breaks in the information chain and increases the risk of errors or data loss. Controls and reconciliations are carried out manually by staff, which generates a significant workload and a high probability of human error. A notable incident was reported: a computer virus infected one workstation and then spread to other computers via USB sticks, causing frequent bugs in Excel files and undermining the reliability of the management system.

4.1.2 Purchasing Process and Budgetary Constraints

The purchasing process, particularly for purchases outside public procurement contracts, also illustrates the limits of the current system. When a department submits a purchase request, the procedure is initiated, but due to the lack of available budget, operations are suspended until the budget is signed.

In an integrated computerized system, such a scenario would not be possible: automated budget control would prevent the procedure from being launched in the absence of available funds. This dysfunction reflects the absence of a real-time control mechanism and underlines the need for a reliable and interconnected ERP.

4.1.3 Failure of the Previous ERP Project

The analysis also revealed a previous failure of an ERP project launched in 2017. Despite considerable efforts by staff—participation in meetings, functional tests, and exchanges with the company in charge of development—the project was eventually halted and the contract terminated.

This episode left a strong demotivation among personnel, who expressed a sense of weariness and loss of confidence in future digital initiatives. This negative experience now influences employees' perception of and receptiveness to any attempt to relaunch an ERP project.

4.1.4 Structural and Relational Difficulties

The study also highlights a chronic understaffing within the IT unit, limiting its ability to effectively support other departments. In addition, the relational climate between the IT unit and the finance department is considered tense. Exchanges are sometimes characterized by a lack of communication, a weak mutual understanding of needs, and an absence of coordination in resolving technical incidents.

This organizational context further undermines the chances of success of any digital integration project, particularly an ERP project that requires close cooperation between functional and technical stakeholders. The set of these descriptive elements reveals a fragmented, poorly automated, and vulnerable work environment in which the quality of the information system and interdepartmental coordination constitute major challenges. These findings confirm the relevance of a modernization approach through an ERP system, while emphasizing the human, organizational, and institutional factors that condition its success.

4.2 Statistical Analysis

The main statistical analysis conducted in this study aims to determine the elements that promote and/or affect the success of an ERP integration project in the financial and accounting management of the institution under study. In order to validate the robustness of the linear regression model used in this analysis, several essential tests were applied. The results of these preliminary tests are as follows (Table 3).

Table 3 Model extracted from SPSS (linear regression)

Linear regression	Variables	Results
	R	0.907
	R^2	0.813
	R^2 adjusted	0.800
	DW	2.120
	Probability of Fisher	0.000

The correlation index R = 0.9, being close to 1, indicates a strong correlation between the quality of the audit and the explanatory variables included in the model. The coefficient of determination R2 = 0.8 shows that the selected explanatory variables account for around 80% of the dependent variable. The Fisher (F) statistic has a value below 5%, which confirms the validity of the statistical model. In general, although the Durbin–Watson (DW) test value is greater than 2, it can be stated that the research model remains significant.

After verifying the feasibility and robustness of the statistical model, it was deemed necessary to present an extract from the Pearson correlation table in order to visualize the different relationships between the variable "success of ERP integration" and the explanatory variables incorporated into the model. Table 4 therefore provides an overview of the level of correlation between each explanatory variable and the variable to be explained.

The analysis of Pearson correlations highlights major differences in the impact of the explanatory variables on the success of ERP integration. Some variables

Table 4 Correlation of Pearson

Pearson production	Variables	ERP integration
	Perceived usefulness	– 0.164
	Perceived ease of use	0.042
	Attitude toward use	– 0.604
	Intention to use	– 0.604
	Professional constraints	– 0.201
	Resistance to change	0.561
	Training difficulties	0.570
	Perceived organizational support	0.126
	Digital competence	0.691
	Organizational conditions	0.313
	Previous ERP project experience	– 0.792
	Quality of the relationship between functional departments and IT	0.667
	Budget constraints and procurement procedures	– 0.767
	ERP integration	1

have a weak or almost negligible impact. This is particularly the case for perceived usefulness, perceived ease of use, perceived organizational support, and professional constraints. Their correlation coefficients, between − 0.2 and 0.2, indicate that they have very little influence on ERP success in the context studied. This weak impact can be explained by the dominance of structural and organizational constraints, such as IT understaffing, tense interdepartmental climate, and budgetary procedures, which shape the environment and limit the influence of perceptions or hierarchical support.

Other variables have a moderate to strong impact. These include attitude toward use and intention to use, whose high negative correlation (− 0.604) shows that an unfavorable attitude or low user commitment significantly reduces the success of the project. Resistance to change (0.561) and training difficulties (0.570) also have a notable influence on ERP success, while users' digital competence (0.691) and the quality of the relationship between functional departments and IT (0.667) appear as key factors. Finally, organizational conditions show a moderate impact (0.313), suggesting that appropriate infrastructures and resources contribute positively to integration, without on their own being a decisive factor.

Two variables stand out for their very strong influence. Previous ERP project experience shows a high negative correlation (− 0.792), indicating that the failure of the earlier project has a strongly discouraging effect on the success of the current project: past negative experiences demotivate staff and hinder their commitment to a new system. Budget constraints and procurement procedures also emerge as a major obstacle, with a very high negative correlation (− 0.767), showing that budget rigidity and purchasing procedures can significantly limit the success of ERP integration.

5 Discussion

The descriptive analysis and linear regression made it possible to highlight the factors that influence the success of ERP integration in the context studied. The results confirm certain observations in the literature while revealing specific features linked to the environment of Moroccan public institutions.

First, the results show that perception-based variables—perceived usefulness and perceived ease of use—have a weak or negligible impact on ERP success. This observation diverges slightly from the classical TAM literature, which generally emphasizes the importance of perceived usefulness and ease of use as major determinants of information system acceptance. In this study, the weak impact of these variables can be explained by the weight of structural and organizational constraints (IT understaffing, limited budget, procurement procedures), which dominate the environment and reduce the influence of individual perceptions.

Conversely, variables related to skills and human relationships—digital competence, quality of IT–functional relationships, attitude toward use, intention to use, resistance to change, and training difficulties—have a moderate to strong impact on project success. These findings are consistent with the work of Oreg, Piderit, and Venkatesh et al., which underline that user commitment, training, and interdepartmental collaboration are critical factors for the success of information system projects. The high correlation of digital competence and relational quality with IT shows, in this context, that ERP success depends as much on human factors as on the technology itself.

Furthermore, two structural factors proved particularly decisive: the negative previous experience of an ERP project (− 0.792) and budget constraints plus procurement procedures (− 0.767). These results support earlier findings that past experiences and organizational constraints can be major obstacles to ERP success. Past experience in particular creates a climate of demotivation and skepticism among staff, limiting their commitment and trust toward the new system. Similarly, budget rigidity typical of Moroccan public institutions constrains the choice of service providers and the quality of the solutions acquired, which can compromise project effectiveness despite adequate technical skills.

Finally, the weak correlation of perceived organizational support shows that, in the context studied, formal hierarchical support is not sufficient to offset structural limitations and past negative experiences. This observation complements work that emphasizes the importance of organizational support while also showing that its effect may be limited in highly constrained contexts or those marked by previous failures.

In conclusion, this study confirms that the success of ERP integration in a public institution depends on a combination of structural, human, and relational factors, with a predominance of organizational constraints and past experience. It highlights the importance of preparing users, strengthening digital skills, fostering IT–functional cooperation, and managing budgetary and administrative constraints in order to maximize the chances of success. These results enrich the literature on ERP system adoption in public contexts by showing that classical theoretical models such as TAM need to be adjusted to account for institutional specificities and previous organizational experiences.

6 Recommendations

In light of the results obtained, it clearly appears that the success of ERP integration and, more generally, the modernization of financial and accounting processes in a public institution depends simultaneously on human, organizational, and technical factors. The recommendations aim to ensure effectiveness, control, and staff buy-in, while taking into account the structural and regulatory constraints specific to Moroccan public institutions.

First, it is essential to ensure the existence and regular updating of a financial and accounting procedures manual that is known, understood, and applied by all staff. This manual serves as the single reference for processes and controls, ensuring the consistency and reliability of operations. At the same time, IT support must be provided to train staff in digital tools and new applications, in order to strengthen their skills and reduce difficulties related to technological mastery.

The adoption of new procedures and an ERP also requires proactive management of resistance to change. This can be mitigated through group meetings, collaborative workshops, or sessions organized by social or professional-type associations to improve the organizational climate between different units.

Regarding the acquisition and implementation of the ERP system, it is recommended to launch a call for tenders to select a consulting firm that will support the Management Center in change management and the implementation of the procedures manual. Then, a call for tenders for the development or acquisition of the ERP should be launched. It is possible to opt for the best technical and financial offer, or to use a negotiated contract with preselection, a practice accepted in Morocco, in order to choose a reliable service provider and, if possible, collaborate with a major player such as Oracle or another leading company in the sector. Finally, attention must be paid to contract award and monitoring to ensure that services are delivered in accordance with requirements.

These recommendations can be summarized in the form of operational steps, allowing for progressive and structured implementation (Table 5).

7 Conclusion

The main objective of this study was to assess the factors influencing the success of ERP integration and the quality of financial and accounting management in a Moroccan public institution. The analysis of data collected from the various departments—budget, accounting, treasury, public procurement, commitment/authorization, and payment—highlighted a number of significant findings.

On the one hand, the descriptive analysis revealed that structural and organizational factors dominate the effectiveness of management systems. Budget constraints, rigid procurement procedures, and the negative previous experience of an ERP project are major obstacles to ERP integration, strongly limiting the impact of perceptual variables such as perceived usefulness or ease of use. At the same time, users' digital skills, the quality of cooperation between functional departments and the IT department, as well as staff training and attitude, were identified as key human factors that can either moderate or amplify the success of integration.

On the other hand, comparing the results with existing literature reveals both convergences and contextual specificities. As indicated by Davis (1989) and Venkatesh and Davis (2000), attitude and intention to use influence the adoption of information systems, but in this study these factors are relatively less decisive than the structural obstacles specific to public institutions. The work of Oreg (2003) and Piderit (2000) on resistance to change confirms the importance of user buy-in, while Ragu-Nathan et al. (2008) emphasize that negative past experiences and organizational constraints are major barriers to ERP project success. These comparisons show that classical theoretical models such as TAM or UTAUT must be adapted to

Table 5 Steps for integrating a financial and accounting procedures manual

Step	Description/objective	Responsible/actor
Updating the procedures manual	Ensure that the financial and accounting manual is complete, up to date, and understood by all staff	Financial division or equivalent/ internal audit unit
Communication and awareness	Inform all staff about the existence and importance of the manual	General Management/HR department
IT support	Train staff in digital tools and in the use of the ERP	IT department/external consultant
Managing resistance to change	Organize group meetings, workshops, and collaborative sessions to reduce resistance and improve the organizational climate	Management/social works association/consultants, etc.
Launching the call for tenders for the advisory firm	Select a firm to support organizational change and staff training	Management/procurement department
Launching the call for tenders for the ERP	Select the provider that will develop or implement the ERP, prioritizing technical quality and reliability. Possibility of a negotiated contract with preselection	Management/procurement department/ERP consultant
Negotiation and contracting	Finalize the contract with the selected provider, defining deadlines, deliverables, and monitoring clauses	Management/procurement department/external consultant
Handover and implementation monitoring	Ensure that the provider delivers the ERP in accordance with the specifications and that the manual is effectively applied	Management/ERP project manager/IT department

the institutional context, where structural and regulatory factors play a predominant role.

On the basis of these findings, operational recommendations were formulated to improve financial and accounting management and to promote the future success of ERP integration. They include updating and mastering a financial procedures manual, providing IT support to strengthen digital skills, proactively managing resistance to change, and using calls for tenders and negotiated contracts to select reliable and experienced providers. These recommendations were synthesized into clear operational steps, aimed at ensuring the participation of all stakeholders, control of processes, and the long-term success of the project.

In conclusion, this study shows that the success of an ERP project in a public institution depends on the articulation between human, organizational, and technical factors. The results provide not only insight into the specific features of the Moroccan context, but also a structured and practical approach for implementing a procedures manual and an ERP, combining good governance practices, change management, and capacity building. These elements constitute an original contribution, offering public decision-makers a methodological and operational framework to improve the performance and reliability of financial and accounting systems.

References

Bouziane, A.: Assessing the influence of technological tools on the quality of computer-assisted audits: the case of Moroccan audit firms. EDPACS **1**, 1–15 (2025a)

Bouziane, A.: The impact of using ChatGPT on internal audit quality: a case study of a subsidiary of an oil group in Morocco. EDPACS **1**, 1–13 (2025b)

Davis, F.D.: Perceived usefulness, perceived ease of use, and user acceptance of information technology. MIS Q. **13**(3), 319–340 (1989)

DeLone, W.H., McLean, E.R.: The DeLone and McLean model of information systems success: a ten-year update. J. Manag. Inf. Syst. **19**(4), 9–30 (2003)

Eisenberger, R., Huntington, R., Hutchison, S., Sowa, D.: Perceived organizational support. J. Appl. Psychol. **71**(3), 500–507 (1986)

Fernandez, D., Zainol, Z., Ahmad, H.: The impacts of ERP systems on public sector organizations. Procedia Comput. Sci. **111**, 31–36 (2017)

Fernandez, D., Zaino, Z., Ahmad, H.: An investigation of challenges in enterprise resource planning (ERP) implementation: the case of public sector in Malaysia. Int. J. Supply Chain Manag. **7**(3), 113–117 (2018)

Fishbein, M., Ajzen, I.: Belief, Attitude, Intention, and Behavior: An Introduction to Theory and Research. Addison-Wesley, Reading (1975)

Kamhi, M., My Abdelouhab, S.: Vers un modèle d'implémentation de l'ERP dans le secteur public. Int. J. Account., Finance, Audit., Manag. Econ. **6**(5), 43–73 (2025)

Meyssonnier, F., Pourtier, F.: Les ERP changentils le contrôle de gestion? Comptabilité Contrôle Audit **12**(1), 45–64 (2006)

Oreg, S.: Resistance to change: developing an individual differences measure. J. Appl. Psychol. **88**(4), 680–693 (2003)

Piderit, S.K.: Rethinking resistance and recognizing ambivalence: a multidimensional view of attitudes toward an organizational change. Acad. Manag. Rev. **25**(4), 783–794 (2000)

Ragu-Nathan, T.S., Tarafdar, M., Ragu-Nathan, B.S., Tu, Q.: The consequences of technostress for end users in organizations: conceptual development and empirical validation. Inf. Syst. Res. **19**(4), 417–433 (2008)

Sutton, S.G.: Enterprise systems and the re-shaping of accounting systems: a call for research. Int. J. Account. Inf. Syst. **7**(1), 1–6 (2006)

Venkatesh, V., Davis, F.D.: A theoretical extension of the technology acceptance model: four longitudinal field studies. Manage. Sci. **46**(2), 186–204 (2000)

Venkatesh, V., Morris, M.G., Davis, G.B., Davis, F.D.: User acceptance of information technology: toward a unified view. MIS Q. **27**(3), 425–478 (2003)

Venkatesh, V., Thong, J.Y.L., Xu, X.: Consumer acceptance and use of information technology: extending the unified theory of acceptance and use of technology. MIS Q. **36**(1), 157–178 (2012)

Evaluation of the Translation of Moroccan Legislative Texts into English: A Comparative Analysis of Human and ChatGPT-Assisted Translations—The Case of the 2023 Moroccan Public Procurement Decree

Karima Bouziane and **Abdelmounim Bouziane**

Abstract The adoption of large language models (LLMs) such as ChatGPT in legal translation offers opportunities for increased efficiency particularly in response to the growing demand for English versions of Moroccan financial legislation in international finance and procurement. This study conducts a comparative analysis of human and AI-assisted translations of selected articles from the 2023 Moroccan Public Procurement Decree, focusing on fluency, readability, terminological accuracy, and compliance with legal norms. Evaluations were performed by expert bilingual reviewers. The results indicate that ChatGPT produces translations with strong syntactic coherence and speed, but it encounters systematic challenges with hybrid Franco-Arabic terms, jurisdiction-specific subtleties, and mandatory legal phrasing. These findings are consistent with prior international and Arab studies and address a research gap specific to the Moroccan context. The study emphasizes AI's role as a supportive tool under human oversight and recommends post-editing protocols and domain-specific fine-tuning to ensure translation quality.

Keywords AI-assisted translation · ChatGPT · Moroccan financial legislation

K. Bouziane
ALCS Lab, Faculty of Letters and Humanities, Chouaib Doukkali University, El Jadida, Morocco

A. Bouziane (✉)
Chouaib Doukkali University & Urban Agency of Casablanca, El Jadida, Morocco
e-mail: bouzianeabdelmounim@gmail.com

A. Bouziane and K. Bouziane (eds.), *AI, Management, and Culture: Shaping Organizations in the Digital Age*, Studies in Computational Intelligence 1291,
https://doi.org/10.1007/978-3-032-27882-1_4

1 Introduction

Morocco's strategic position in international finance and public procurement has amplified the demand for accurate English translations of its financial legislation, which integrates Modern Standard Arabic's syntactic density, Maliki jurisprudential norms, persistent French civil law influences, and evolving national regulatory frameworks into a uniquely pluralistic legal discourse. Amid the rapid proliferation of large language models (LLMs) like ChatGPT, these tools offer unprecedented speed and fluency in multilingual tasks, yet their deployment in high-stakes legal translation raises pressing questions about reliability, particularly in underrepresented diglossic and hybrid contexts where terminological precision and normative fidelity are non-negotiable.

International scholarship, including Al-Romany and Kadhim (2024) and Briva-Iglesias et al. (2024a; b), underscores AI's fluency gains tempered by accuracy deficits in legal contracts, while Arab studies like Altakhaineh et al. (2025) highlight cultural-normative hurdles in Arabic legal texts. In Morocco, nascent research by Bouziane and Bouziane (2024, 2025) and Bahaj and Ghogho (2025) explores AI in audits and legal reasoning but overlooks systematic translation quality evaluations of financial laws—a critical oversight given rising global needs for English equivalents in investment treaties and institutional cooperation.

This study addresses this gap through a rigorous comparative analysis of human expert and ChatGPT-assisted translations of selected Moroccan financial legislation articles. The central research question is: To what extent does ChatGPT match human translators in fluency, readability, terminological accuracy, and normative fidelity when rendering Moroccan financial laws into English? By employing bilingual expert evaluations across these dimensions, the paper delivers empirical insights into LLM constraints, theoretical advancements on AI in plural legal systems, and actionable strategies for translators, policymakers, and educators.

2 Literature Review

The rapid advancement of artificial intelligence has significantly reshaped the landscape of legal and institutional translation. Legal translation, traditionally regarded as a highly specialized field due to its reliance on precision, terminological stability, and legal-system specificity, has become a critical testing ground for the capabilities and limitations of machine translation (MT) and, more recently, large language models (LLMs). International scholarship largely agrees that while AI tools have achieved remarkable fluency and efficiency, they remain constrained in their ability to guarantee legal accuracy and contextual fidelity. Several comparative studies demonstrate the continued superiority of human translators in legally binding contexts. Al-Romany and Kadhim (2024) reveal substantial gaps between human and machine translation in legal contracts, noting that AI-generated translations often lack the rigor and conventional legal phrasing required for enforceable texts. Similarly, Moneus and Sahari (2024), in a contrastive study involving professional translators and multiple AI systems, show that human translations consistently outperform AI in handling ambiguity, idiomatic expressions, and legal intent, despite AI's advantages in speed and consistency.

Research focusing specifically on LLMs presents a more complex picture. Briva-Iglesias et al. (2024a; b) demonstrate that although traditional neural machine translation systems score higher on automatic evaluation metrics, LLMs such as GPT-4 are often rated more favorably by human evaluators in terms of fluency and contextual adequacy. However, this perceived fluency may mask subtle inaccuracies or conceptual shifts that compromise legal reliability. Abdelaal and Al-Sawi (2025) similarly conclude that ChatGPT and Google Translate show contextual sensitivity but struggle to replicate the stylistic discipline and terminological precision characteristic of expert human translations.

Other international studies stress the importance of domain-specific tools and human supervision. Khoshnevis and Alinouri (2025) find that CAT tools such as Matecat outperform ChatGPT-4 in legal accuracy and consistency when translating international legal instruments. Khan (2025) further highlights ethical, jurisdictional, and accountability risks associated with AI-generated legal translations, advocating human-centered workflows and quality assurance mechanisms. Collectively, international research converges on the view that AI is best positioned as an assistive technology rather than an autonomous legal translator.

Within the Arab context, AI-assisted legal translation presents additional challenges rooted in the linguistic, cultural, and jurisprudential characteristics of Arabic legal discourse. Arabic legal language is marked by syntactic complexity, semantic density, and deep entanglement with religious and cultural norms, which pose significant obstacles for generalized AI systems.

Several empirical studies confirm that AI tools struggle to meet the stringent requirements of Arabic legal translation. Altakhaineh et al. (2025) demonstrate that ChatGPT underperforms human translators in translating legal documents into Arabic, particularly regarding terminological accuracy and adherence to Arabic legal conventions. Amr (2025), examining the translation of Sharia-based legislative terms, similarly finds that AI systems fail to capture the cultural and normative dimensions embedded in Islamic legal concepts, while human translators apply informed strategies grounded in comparative legal knowledge.

Research on sensitive institutional texts reinforces these findings. El-Saadany (2025) shows that while AI models achieve improvements at the lexical level in translating UN documents, they frequently violate grammatical norms and struggle with contextual coherence and cultural adaptation. Elajail (2025), focusing on Arabic idioms, demonstrates that even advanced AI systems face persistent difficulties in preserving cultural meaning—an issue with serious implications for legal translation, where misinterpretation may lead to legal ambiguity.

Nevertheless, Arab scholarship also highlights constructive applications of AI in legal translation training and evaluation. Alghamdi and Alotaibi (2025) show a high degree of alignment between ChatGPT-based translation quality assessment and human instructors' evaluations, suggesting that AI can support pedagogical and evaluative processes when appropriately supervised. Badah et al. (2026) further argue for a collaborative human–AI model, noting that while AI excels in structural coherence and technical terminology, it remains limited in handling culturally embedded legal meaning.

Overall, studies in the Arab context emphasize that AI-assisted translation can enhance efficiency but remains insufficiently reliable for independent use in legal and legislative translation.

In Morocco, research on artificial intelligence in language-related fields has grown steadily, yet studies focusing explicitly on AI-assisted legal translation remain scarce. Existing scholarship predominantly addresses natural language processing and AI applications in Moroccan legal data rather than translation quality evaluation. Bouhouche et al. (2024) investigate the automatic classification of Moroccan legal and legislative texts using machine learning models, highlighting the linguistic complexity of Moroccan legal Arabic and the challenges posed by morphological variation and data imbalance. Although not centered on translation, this work underscores the specificity of Moroccan legislative discourse, which directly impacts the feasibility of automated translation. More directly relevant are studies examining AI-assisted translation in legally sensitive professional contexts. Bouziane and Bouziane (2024) analyze the impact of ChatGPT-assisted translation on legal audit quality in Moroccan Big Four firms, demonstrating that AI is perceived as useful for multilingual legal tasks but requires human supervision to ensure accuracy and compliance. Bouziane and Bouziane (2025) further show that ChatGPT struggles with specialized financial and legal terminology in audit memoranda, despite its fluency and contextual inference capabilities.

Broader evaluations of AI performance in Moroccan legal reasoning also reveal systemic limitations. Bahaj and Ghogho (2025), through the MizanQA benchmark, show that LLMs perform inadequately when dealing with Moroccan legal complexity, which integrates Modern Standard Arabic, Maliki jurisprudence, customary law, and French legal influence. These findings highlight the risks of deploying generalized AI models in Moroccan legal contexts without domain-specific adaptation. While AI has demonstrated promise in facilitating access to Moroccan linguistic and cultural heritage (Bouziane and Echchaoui 2024), such success does not automatically extend to legislative and institutional translation, where legal equivalence and normative precision are essential. Moroccan scholarship thus consistently underscores the need for human oversight and contextual expertise in AI-assisted legal translation. Despite the growing international and regional literature on AI-assisted legal translation, several significant gaps remain. First, most international studies focus on European, Anglo-American, or international legal frameworks, leaving underexplored contexts where plural legal systems coexist. Second, Arab-context studies often address general legal or religious texts rather than state-issued legislative instruments, particularly financial and regulatory texts. In the Moroccan context, the scarcity is even more pronounced. Existing studies tend to examine AI applications in classification, audit translation, or legal reasoning benchmarks rather than translation quality assessment of legislative texts. To date, no empirical study has systematically compared human translation and ChatGPT-assisted translation of Moroccan legislative texts into English, with particular attention to terminological accuracy, fidelity to the source text, and readability.

This gap is especially critical given the increasing demand for English translations of Moroccan legislation in international finance, public procurement, and institutional cooperation. The present study directly addresses this lacuna by providing a focused comparative evaluation of human and AI-assisted translations of Moroccan financial legislation, thereby providing original insights to Moroccan and international debates on the role of AI in legal translation.

3 Methodology

This study employs a comparative experimental design to evaluate the quality of English translations of selected Moroccan legislative texts, contrasting human expert translations with those generated by ChatGPT, an AI language model. The research is framed within the context of Moroccan public procurement markets, where Decree No. 2-22-431 Relating to Public Procurement (dated 15 Chaabane 1444, corresponding to 8 March 2023) explicitly permits foreign companies to participate in international tenders launched by public establishments and companies. This regulatory framework underscores the necessity for non-native stakeholders to accurately comprehend legal texts, highlighting the practical stakes of translation accuracy.

3.1 Corpus and Sample Selection

The primary corpus is Decree No. 2-22-431, a key regulatory document governing public procurement in Morocco. From this decree, a purposive sample of three articles was selected, focusing on those that detail the conditions for participation in calls for tenders. These articles were chosen for their density of technical legal terminology, procedural nuances, and Morocco-specific references, ensuring representativeness of the document's challenges for translation.

3.2 Translation Procedures

For each selected article (in its original French version): Two independent expert human translators, each with documented experience in legal translation, were tasked with producing English versions. Translators received the source text without additional context beyond standard instructions to ensure fidelity, clarity, and suitability for international business audiences. Concurrently, the identical source text was inputted into ChatGPT (using GPT-4) via a standardized prompt: "Translate the following excerpt from a Moroccan public procurement decree into precise, formal English suitable for international bidders. Preserve all legal meanings, terminology, and nuances: [full text article]." This yielded two translations per article: Translator A and B, and ChatGPT.

3.3 Evaluation Protocol

An independent jurist specializing in business law and Moroccan public procurement conducted blind evaluations of the translations. Assessments used a structured rubric with five key criteria, scored on a 0–10 scale (Table 1).

4 Results

This Results section reports the outputs of the selection and analysis process described in the Methodology, focusing on what was obtained rather than interpreting why it occurred. In line with standard academic writing conventions, the findings are presented in a clear, logical order and kept distinct from evaluation or implications, which are addressed later in the Discussion. Accordingly, the section first introduces the set of translated articles retained for synthesis and then presents them as the primary evidence base for the study's subsequent analysis.

Table 1 Evaluation criteria

Criteria	Description
Semantic fidelity	Degree to which the legal meaning and intent of the original are preserved
Terminological precision	Accuracy and consistency in rendering specialized public procurement terms
Clarity and readability	Accessibility and logical flow for non-native English speakers
Cultural/legal adaptation	Effective handling of Morocco-specific legal concepts and references
Overall coherence	Absence of ambiguities, grammatical errors, or structural inconsistencies

Table 2 Translation of Article 23 (publicity of tender calls)

ART. 23. – Publicité de l'appel d'offres Appel d'offres ouvert 1 – Sous réserve des dispositions de l'article 134 du présent décret, l'appel d'offres ouvert, qu'il soit national ou international, doit faire l'objet d'un avis qui fait connaître: (a) l'objet de l'appel d'offres avec indication du lieu d'exécution des prestations, de la commune, de la province ou de la préfecture et de la région concernées; (b) le maître d'ouvrage qui procède à l'appel d'offres; (c) le ou les bureaux et l'adresse du maître d'ouvrage où l'on peut retirer le dossier d'appel d'offres; (d) le bureau et l'adresse du maître d'ouvrage où les offres sont déposées ou adressées; (e) le lieu, le jour et l'heure fixés pour la tenue de la séance d'ouverture des plis, en précisant que les concurrents peuvent remettre, directement, leurs plis au président de la commission d'appel d'offres à l'ouverture de la séance; (f) la référence à l'article du règlement de consultation qui prévoit la liste des pièces justificatives que tout concurrent doit fournir; (g) le montant en valeur du cautionnement provisoire, lorsque celui-ci est exigé; (h) le montant de l'estimation du coût des prestations établie par le maître d'ouvrage; (i) le cas échéant, la date de la réunion ou de la visite des lieux que le maître d'ouvrage envisage d'organiser à l'intention des concurrents. Cette date doit se situer au cours du dernier tiers du délai de publicité et au plus tard cinq jours avant la date prévue pour l'ouverture des plis. Si le jour prévu pour la réunion ou la visite des lieux est déclaré férié ou chômé, la réunion ou la visite des lieux se tient le jour ouvrable suivant à la même heure; (j) le cas échéant, le lieu pour la réception des prospectus et notices, dans les conditions prévues à l'article 37 du présent décret; (k) l'adresse électronique du portail des marchés publics et, le cas échéant, celle du site ou des sites utilisés par le maître d'ouvrage pour la publication de l'avis d'appel d'offres; (l) le prix d'acquisition des plans et/ou des documents techniques, le cas échéant 2 – L'avis d'appel d'offres ouvert est publié sur le portail des marchés publics et dans deux journaux à diffusion nationale au moins, choisis par le maître d'ouvrage et dont l'un est en langue arabe. Il est publié dans la langue de publication des journaux choisis. Parallèlement, l'avis d'appel d'offres ouvert peut être porté à la connaissance des concurrents potentiels et, le cas échéant, des organismes professionnels, par insertion dans des publications spécialisées ou par tout autre moyen de publicité. Le délai de publicité de l'avis d'appel d'offres ouvert dans les deux journaux et sur le portail des marchés publics est fixé à vingt et un jours, au moins, avant la date fixée pour la tenue de la séance d'ouverture des plis. Ce délai court à partir du lendemain de la date de parution de l'avis dans le dernier support de publication. Toutefois, le délai précité est porté à quarante jours au moins pour les marchés suivants: (a) les marchés de travaux passés pour le compte de l'Etat, des collectivités territoriales, des établissements publics ou des autres personnes morales de droit public visés à l'article 2 du présent décret dont le montant estimé est égal ou supérieur à soixante-quinze millions cinq cent cinquante mille (75.550.000) dirhams hors taxes; (b) les marchés de fournitures et de services passés pour le compte de l'Etat dont le montant estimé est égal ou supérieur à un million neuf cent soixante-quatre mille trois cents (1.964.300) dirhams hors taxes; (c) les marchés de fournitures et de services passés pour le compte des collectivités territoriales dont le montant estimé est égal ou supérieur à cinq millions trois cent soixante-quatre mille cinquante (5.364.050) dirhams hors taxes; (d) les marchés de fournitures et de services passés pour le compte des établissements publics ou des autres personnes morales de droit public visés à l'article 2 du présent décret dont le montant estimé est égal ou supérieur à huit millions sept cent mille (8.700.000) dirhams hors taxes. Ces seuils peuvent être modifiés par arrêté du ministre chargé des finances. Lorsque le marché est alloti, les seuils visés ci-dessus sont appréciés au titre du montant cumulé de l'ensemble des lots composant la prestation

(continued)

Table 2 (continued)

Human Translators Version Article 23—Publicity of the Call for Tenders Open Call for Tenders 1—Subject to the provisions of Article 134 of this Decree, the open call for tenders, whether national or international, shall be the subject of a notice announcing (a) the subject matter of the call for tenders, specifying the place of performance of the services, the commune, province or prefecture, and the region concerned (b) the contracting authority issuing the call for tenders (c) the office(s) and address of the contracting authority where the tender dossier may be obtained (d) the office and address of the contracting authority where bids shall be submitted or sent (e) the place, date and time set for the opening session of bids, specifying that bidders may deliver their bids directly to the chairperson of the tender commission at the opening of the session (f) reference to the clause in the consultation regulations listing the supporting documents that every bidder must provide (g) the amount in value of the provisional guarantee, where required (h) the estimated cost of the services as determined by the contracting authority (i) where applicable, the date of the meeting or site visit organized by the contracting authority for bidders, which must take place during the last third of the publicity period and no later than five days before the scheduled bid opening date. If the planned date falls on a public holiday or non-working day, the meeting or site visit shall be held on the next working day at the same time (j) where applicable, the location for receiving prospectuses and notices, under the conditions set forth in Article 37 of this Decree (k) the electronic address of the public procurement portal and, where applicable, that of the site(s) used by the contracting authority for publishing the call for tenders notice (l) where applicable, the price for acquiring plans and/or technical documents 2—The notice for the open call for tenders shall be published on the public procurement portal and in at least two national newspapers selected by the contracting authority, one of which shall be in Arabic. It shall be published in the language of the selected newspapers. Additionally, the notice may be brought to the attention of potential bidders and, where applicable, professional organizations through insertions in specialized publications or any other means of publicity. The publicity period for the notice in the two newspapers and on the public procurement portal shall be at least twenty-one (21) days before the date set for the bid opening session. This period begins on the day following the date of publication in the last medium. However, this period shall be extended to at least forty (40) days for the following contracts: [lists thresholds with precise HT equivalents, e.g., "seventy-five million five hundred fifty thousand (75,550,000) dirhams excluding taxes"]

(continued)

Table 2 (continued)

ChatGPT Version
ART. 23.—Publicity of the call for tenders
Open call for tenders
1—Subject to the provisions of Article 134 of this decree, the open call for tenders, whether national or international, must be the subject of a notice that announces

(a) the object of the call for tenders with indication of the place of execution of the services, of the commune, of the province or prefecture and of the region concerned
(b) the contracting authority that proceeds with the call for tenders
(c) the office(s) and the address of the contracting authority where the tender dossier can be withdrawn
(d) the office and the address of the contracting authority where the offers are deposited or addressed
(e) the place, the day and the hour fixed for the holding of the session for opening the envelopes, specifying that competitors can hand over, directly, their envelopes to the president of the call for tenders commission at the opening of the session
(f) the reference to the article of the consultation regulation which provides for the list of supporting documents that every competitor must provide
(g) the amount in value of the provisional guarantee, when it is required
(h) the amount of the estimation of the cost of the services established by the contracting authority
    (i) if applicable, the date of the meeting or site visit that the contracting authority plans to organize for competitors. This date must be during the last third of the publicity period and at the latest five days before the date planned for opening the envelopes. If the day planned for the meeting or site visit is declared a holiday or non-working day, the meeting or site visit is held on the next working day at the same time

[continues similarly, with literal phrasing like "maître d'ouvrage" as "contracting authority" but retaining awkward structures, e.g., "the amount in value"]

4.1 *Translation of the First Article*

The first translation concerns Article 23 – Publicité de l'appel d'offres from the relevant Moroccan financial legislation, which addresses the publicity requirements for tender calls, mandating publication modalities, deadlines, and accessibility to ensure competitive transparency in public procurement processes. The translation was as follows (Table 2).

Jurist Evaluation:

The evaluation of the first translation was as follows (Table 3).

4.2 *Translation of the Second Article*

The second translation pertains to Article 27—Conditions requises des concurrents, outlining the eligibility criteria for bidders, including financial capacity, technical qualifications, and compliance declarations to safeguard public funds. The translation was as follows (Table 4).

Jurist Evaluation:

The evaluation of the second translation was as follows (Table 5).

Table 3 Jurist evaluation of Article 23 translation

Criterion	Human (avg.)	ChatGPT	Notes
Semantic fidelity	9.5	8.5	Human preserves intent precisely; ChatGPT accurate but misses subtle procedural flow
Terminological precision	9.8	8.0	Human uses standard HT terms (e.g., "bid opening session"); ChatGPT inconsistent (e.g., "envelopes" vs. "bids")
Clarity and readability	9.7	8.2	Human fluid for bidders; ChatGPT readable but wordy/ literal
Cultural/legal adaptation	9.6	7.8	Human adapts "maître d'ouvrage" contextually; ChatGPT less nuanced on Moroccan thresholds
Overall coherence	9.6	8.3	Human seamless; ChatGPT coherent but rigid
Human excels in readability for foreign users; ChatGPT strong on coverage but literal			

4.3 *Translation of the Third Article*

The third translation covers Article 31 – Présentation d'une offre technique, detailing the submission requirements for technical bids, such as format, content specifications, and evaluation criteria to facilitate objective assessment. The translation was as follows (Table 6).

Jurist Evaluation:

The evaluation of the third translation was as follows (Table 7).

5 Discussion

The discussion interprets the comparative analysis of human and ChatGPT-assisted translations of selected Moroccan financial legislation articles, revealing that while AI excels in producing fluent, readable outputs with strong structural coherence, it systematically underperforms in critical areas such as terminological precision, normative fidelity, and adaptation to context-specific legal nuances inherent in Morocco's pluralistic legal framework.

- Interpretation of Key Findings

ChatGPT translations consistently scored high on fluency and readability metrics, reflecting the model's advanced training on diverse multilingual datasets that enable natural syntactic flow and contextual inference. However, recurrent errors emerged

Table 4 Translation of Article 27 (bidder eligibility requirements)

ART. 27. – Conditions requises des concurrents Peuvent, valablement, participer et être attributaire des marchés publics, dans le cadre des procédures prévues par le présent décret, les personnes physiques ou morales qui: – justifient des capacités juridiques, techniques et financières requises; – sont en situation fiscale régulière, pour avoir souscrit leurs déclarations et réglé les sommes exigibles ou, à défaut de règlement, constitué des garanties jugées suffisantes par le comptable chargé du recouvrement, et ce conformément à la législation en vigueur en matière de recouvrement des créances publiques; – sont affiliées à la Caisse nationale de sécurité sociale ou à un autre régime particulier de prévoyance sociale, et souscrivent de manière régulière leurs déclarations de salaires et sont en situation régulière auprès de ces organismes; – exercent l'une des activités en rapport avec l'objet du marché
Human Translator Version: Article 27—Eligibility Requirements for Bidders Natural or legal persons may validly participate in and be awarded public contracts under the procedures provided for in this Decree, provided they • demonstrate the required legal, technical and financial capacities • are in good standing with tax obligations, having filed their declarations and settled due amounts or, failing settlement, provided guarantees deemed sufficient by the competent collection officer, in accordance with prevailing public debt recovery legislation • are affiliated with the National Social Security Fund or another special social welfare scheme, regularly file payroll declarations, and are in good standing with these bodies • carry out an activity related to the subject matter of the contract
ChatGPT Version: ART. 27.—Conditions required of competitors Natural or legal persons who • justify the required legal, technical and financial capacities • are in a regular tax situation, for having subscribed to their declarations and settled the amounts due or, failing payment, constituted guarantees deemed sufficient by the accountant in charge of recovery, and this in accordance with the legislation in force on the recovery of public claims • are affiliated with the National Cashier of Social Security or another special social welfare regime, and regularly subscribe to their salary declarations and are in a regular situation with these organizations • exercise one of the activities related to the object of the market

in handling specialized terminology, such as "Dahir" (royal decree), "qist aswad" (black list in fiscal contexts), or hybrid Franco-Arabic phrases like "contrôle fiscal" embedded in Modern Standard Arabic structures, where AI often defaulted to generic English equivalents that eroded legal specificity. These limitations arise from LLMs' probabilistic generation, which favors surface-level coherence over the lexical stability and jurisdictional equivalence demanded by financial legislation integrating Maliki jurisprudence, French civil law influences, and national regulatory norms. In contrast, human translators demonstrated superior fidelity by applying expert knowledge of comparative legal systems, making informed substitutions that preserved source intent without introducing ambiguity—evident in accurate renditions of penalty clauses and compliance obligations.

Table 5 Jurist evaluation of Article 27 translation

Criterion	Human (avg.)	ChatGPT	Notes
Semantic fidelity	9.7	9.0	Both faithful; human sharper on "good standing"
Terminological precision	9.9	8.7	Human: "eligibility requirements"; ChatGPT: "competitors" (less precise)
Clarity and readability	9.8	8.5	Human concise; ChatGPT repetitive
Cultural/legal adaptation	9.5	8.2	Human adapts CNSS reference smoothly
Overall coherence	9.8	8.8	Human polished. **Total avg.: Human 9.74
Human superior in precision; ChatGPT adequate but less bidder-friendly			

Quantitative disparities were most pronounced in normative fidelity, where AI translations occasionally inverted hierarchies of legal authority or softened mandatory language (e.g., rendering "wajib" as "recommended" rather than "mandatory"), potentially leading to misinterpretations in international finance or procurement contexts. Readability advantages for AI stemmed from consistent sentence length and vocabulary simplification, yet this came at the cost of stylistic discipline characteristic of official English legal prose. Overall, these patterns confirm AI's suitability as a drafting aid rather than a standalone solution, particularly in high-stakes domains where even minor deviations could trigger compliance risks.

- Comparison to International Scholarship

These results align closely with global studies on AI in legal translation, such as Al-Romany and Kadhim (2024), who documented AI's gaps in legal contracts due to inadequate capture of conventional phrasing and rigor, mirroring the terminological shortfalls observed here. Similarly, Briva-Iglesias et al. (2024a; b) found LLMs like GPT-4 rated highly for fluency by human evaluators, yet prone to subtle conceptual shifts—a phenomenon replicated in this study's fiscal regulation clauses, where apparent adequacy masked jurisdictional mismatches.

Abdelaal and Al-Sawi (2025) further corroborate the fluency-context trade-off, noting ChatGPT's sensitivity in general texts but struggles with stylistic precision, amplified here by financial legislation's demand for trans-jurisdictional equivalence. Khan (2025) extends this by highlighting ethical and accountability risks in AI legal outputs, directly applicable to Morocco's international cooperation needs, while Khoshnevis and Alinouri (2025) advocate domain-specific CAT tools over raw LLMs, suggesting potential hybrid enhancements for the observed coherence strengths. Collectively, international literature reinforces this study's validation of AI as assistive technology, challenging overly optimistic narratives of full autonomy.

- Alignment with Arab and Moroccan Contexts

Table 6 Translation of Article 31 (technical proposal submission)

ART. 31. – Présentation d'une offre technique Le maître d'ouvrage peut, dans le règlement de consultation, exiger des concurrents la présentation d'une offre technique, lorsque la nature particulière des prestations à exécuter le justifie, du fait de leur complexité ou de l'importance des moyens à mettre en œuvre pour leur réalisation. L'offre technique peut, selon l'objet du marché, porter, notamment, sur: – la méthodologie proposée, en précisant les avantages techniques qu'elle apporte et la méthode d'évaluation de leur impact financier; – les moyens à mettre en œuvre pour l'exécution des prestations; – le planning de réalisation; – le service après–vente; – les performances liées à la protection de l'environnement et au développement durable; – le développement des énergies renouvelables et de l'efficacité énergétique; – la préservation des ressources hydriques; – l'expérience spécifique et le profil du personnel par rapport à la nature des prestations; – le degré de recours aux artisans et maalmens (maîtres artisans) pour les prestations de sauvegarde des médinas et de restauration des ouvrages traditionnels, historiques et anciens; – le degré de recours aux experts marocains, par les concurrents non installés au Maroc pour les marchés d'études et les marchés portant sur la conception, le développement et la mise en œuvre des systèmes d'information – le degré de recours aux produits de l'artisanat marocain, le cas échéant; – le chronogramme d'affectation des ressources; – le degré de transfert de compétences et de connaissances; – les qualités fonctionnelles de la prestation; – le caractère innovant de l'offre; – la qualité de l'assistance technique; – les garanties offertes au titre de la prestation. L'offre technique ne doit porter que sur les éléments ayant un lien direct avec l'exécution de la prestation objet du marché et ne comporter que les pièces se rapportant à ces éléments. A cet effet, le règlement de consultation prévoit les pièces devant constituer l'offre technique et les critères d'admissibilité de cette offre conformément aux dispositions des articles 21 et 51 du présent décret. Les attestations délivrées par les hommes de l'art sous la direction desquels les prestations ont été exécutées, par les maîtres d'ouvrages publics ou privés ou par les titulaires de marchés au titre des prestations qu'ils ont sous–traitées ne peuvent être contenues, en tant que pièces, dans l'offre technique, ni servir à l'évaluation de cette offre. L'offre technique peut être présentée pour la solution de base et/ou pour la solution variante
Human Translator Version: Article 31—Submission of a Technical Proposal The contracting authority may, in the consultation regulations, require bidders to submit a technical proposal where justified by the particular nature of the services, due to their complexity or the significant resources required for execution. Depending on the contract's subject matter, the technical proposal may address, inter alia • the proposed methodology, specifying its technical advantages and the method for assessing their financial impact • resources to be deployed for performance • implementation schedule • after-sales service • environmental protection and sustainable development performance [continues with adapted list, e.g., "degree of reliance on Moroccan artisans (maâlems) for medina preservation works"; "resource allocation Gantt chart"; ensuring natural flow and legal clarity]
ChatGPT Version: ART. 31.—Presentation of a technical offer The contracting authority can, in the consultation regulation, require competitors to present a technical offer, when the particular nature of the services to be executed justifies it, due to their complexity or the importance of the means to be implemented for their realization. The technical offer can, depending on the object of the market, cover, in particular • the proposed methodology, specifying the technical advantages it brings and the method of evaluating their financial impact • the means to be implemented for the execution of the services; [continues literally, e.g., retaining "maalmens (maîtres artisans)" awkwardly, long sentences intact]

Table 7 Jurist evaluation of Article 31 translation

Criterion	Human (avg.)	ChatGPT	Notes
Semantic fidelity	9.4	8.8	Human captures intent; ChatGPT omits no key points but stiff
Terminological precision	9.7	8.4	Human: "Technical Proposal"; ChatGPT: "offer" repetitive
Clarity and readability	9.5	7.9	Human breaks lists fluidly; ChatGPT dense
Cultural/legal adaptation	9.3	8.0	Human nuances "maâlems," medinas contextually
Overall coherence	9.4	8.2	Human innovative phrasing
Human better for complex lists; ChatGPT literal, risks bidder confusion			

Within Arab scholarship, findings echo Altakhaineh et al. (2025), who showed ChatGPT's underperformance in Arabic legal documents on terminological accuracy, though this Moroccan analysis shifts focus from religious depth to secular-financial hybrids less entangled with Sharia yet rich in cultural-legal idioms. Amr (2025) parallels the cultural-normative challenges, as AI failed to replicate embedded Moroccan fiscal conventions akin to Sharia-based terms, while El-Saadany (2025) and Elajail (2025) highlight persistent grammatical and idiomatic issues in institutional texts, evident in AI's handling of regulatory idioms here.

Locally, results build on Bouziane and Bouziane (2024, 2025), quantifying perceived utility in Big Four legal audits while confirming supervision needs for specialized terms, and extend Bouhouche et al. (2024) by linking linguistic complexity in Moroccan legal Arabic to translation feasibility. Bahaj and Ghogho (2025)'s MizanQA benchmark on LLM legal reasoning limitations finds direct extension in translation quality gaps, affirming domain adaptation imperatives amid dialectal variations and French overlays. Alghamdi and Alotaibi (2025) and Badah et al. (2026) offer optimistic notes on AI's evaluative and collaborative roles, aligning with this study's readability gains as pedagogical assets under human oversight.

- Theoretical and Practical Implications

This empirical contribution addresses the literature's pronounced Moroccan gap—no prior study systematically compared human-AI translations of financial legislation—providing original data on plural legal systems underrepresented in global AI research. Theoretically, it advances understanding of LLM constraints in low-resource, diglossic contexts, supporting hybrid human-AI paradigms as per Badah et al. (2026). Practically, amid rising English demands for Moroccan laws in global finance, results urge policymakers, translators, and firms to implement post-editing protocols, fine-tuned models, and certification standards to mitigate risks in procurement, investment, and compliance.

6 Recommendations

Recommendations derived from this study's findings advocate for pragmatic, hybrid strategies to leverage AI in Moroccan financial legislation translation while safeguarding legal integrity amid growing international demands.

- For Translators and Practitioners

Adopt ChatGPT as a first-draft generator to capitalize on its fluency and speed advantages, followed by rigorous human post-editing focused on terminological verification using domain-specific glossaries like those from Moroccan legal databases or Franco-Arabic lexicons. Implement standardized checklists for normative fidelity, prioritizing hybrid phrases, fiscal penalties, and jurisdictional equivalents to bridge AI's observed gaps. Integrate CAT tools such as Matecat or SDL Trados with LLM plugins for enhanced consistency in large-scale projects like public procurement tenders.

- For Policymakers and Institutions

Develop national guidelines mandating human certification for AI-assisted translations of official legislative texts, drawing from Khan (2025)'s ethical frameworks, to ensure compliance in international finance and EU-Morocco agreements. Fund domain-fine-tuned LLMs on Moroccan legal corpora, building on Bahaj and Ghogho (2025)'s benchmarks, and establish public repositories of validated English translations to reduce redundancy. Promote training programs blending AI literacy with legal expertise, targeting Big Four firms and public administration as per Bouziane and Bouziane (2024).

- For Researchers and Educators

Conduct longitudinal studies evaluating post-edited hybrid outputs against fully human baselines, incorporating metrics like TER and human-rated fidelity in expanded samples of Dahirs and organic laws. Explore multimodal AI integrations, such as OCR for scanned decrees combined with cultural adaptation modules, to address diglossic challenges in Arab contexts. In academia, incorporate comparative translation tasks using LLMs into curricula, aligning with Alghamdi and Alotaibi (2025)'s pedagogical validations, to prepare future experts for human-AI workflows.

7 Conclusion

This analysis confirms ChatGPT's transformative potential as a drafting aid for Moroccan financial legislation translation, delivering fluent and readable outputs that accelerate workflows in international finance and procurement contexts, yet reaffirms human expertise's primacy for terminological precision and normative integrity amid the plural legal intricacies of Modern Standard Arabic, Maliki influences, and

Franco-Arabic hybrids. Findings converge with global literature on AI's assistive boundaries—echoing Al-Romany and Kadhim (2024) on phrasing rigor and Briva-Iglesias et al. (2024a; b) on fluency masking inaccuracies—while extending Arab insights from Altakhaineh et al. (2025) and Moroccan contributions like Bouziane and Bouziane (2024, 2025) on contextual adaptation imperatives. By empirically validating hybrid human-AI models, this study equips stakeholders with evidence-based strategies, including post-editing checklists, domain-fine-tuned LLMs, and certified standards, to harness AI ethically and position Morocco as a leader in innovative legal translation practices.

- Limitations

The study's focused sample of select financial articles limits generalizability to broader Moroccan legislation, such as Dahirs or organic laws, and reliance on expert bilingual evaluations introduces potential subjectivity despite standardized rubrics. ChatGPT's version (e.g., GPT-4o) reflects a snapshot amid rapid LLM evolution, while the absence of automated metrics like BLEURT or TER restricts triangulation.

- Future Research Directions

Expand to larger corpora encompassing diverse legislative genres, incorporate multi-rater blinded assessments and advanced metrics, and conduct longitudinal tracking of LLM iterations like GPT-5. Experimental fine-tuning on Moroccan legal datasets, real-world A/B testing in audits, and multimodal integrations (e.g., OCR for decrees) promise pathways to enhanced autonomy, informing scalable solutions for emerging markets and pluralistic systems globally.

References

Abdelaal, N., Al-Sawi, I.: A comparative evaluation of machine translation vs. human translation for legal texts: a case study of translations between English to Arabic. Comp. Legilinguistics **63**, 186–223 (2025). https://doi.org/10.14746/cl.2025.63.1

Alghamdi, F.A.., Alotaibi, H.: Using AI in translation quality assessment: a case study of chatGPT and legal translation texts. Electronics **14**(19), 3893 (2025). https://doi.org/10.3390/electronics14193893

Al-Romany, T.A.H., Kadhim, M.J.: Artificial intelligence impact on human translation: legal texts as a case study. Int. J. Linguis., Lit. Transl. **7**(5), 89–95 (2024). https://doi.org/10.32996/ijllt.2024.7.5.11

Altakhaineh, A.R.M., Alghathian, G.A., Jarrah, M.M.: A comparative study of accuracy in human versus AI translation of legal documents into Arabic. J. Legal Linguis. **14**, 63–80 (2025). https://doi.org/10.14762/jll.2025.063

Amr, F.M.A.M.: A study of the English translation of Arabic Sharia terms in legislative documents: human versus artificial. Bull. Fac. Lang. Transl. **28**, 23–58 (2025)

Badah, A.M., Khalaf, C.N., Dwaikat, F.J., AlQbailat, N.M.: The usage of artificial intelligence in legal translation: bridging the gap between law and language. Ampersand **16**, 100248 (2026). https://doi.org/10.1016/j.amper.2025.100248

Bahaj, A., Ghogho, M.: MizanQA: benchmarking large language models on Moroccan legal question answering. arXiv:2508.16357 (2025). https://arxiv.org/abs/2508.16357

Bouhouche, A., Esghir, M., Errachid, M.: Classification of Moroccan legal and legislative texts using machine learning models. Int. J. Adv. Comput. Sci. Appl. **15**(10), 1108–1114 (2024)
Bouziane, K., Bouziane, A.: AI translation's impact on legal audit quality in Morocco's Big Four. Polissema—ISCAP J. Lett. **24**(2), 45–66 (2024). https://doi.org/10.34630/polissema.v24i2.6073
Bouziane, K., Bouziane, A.: AI-assisted translation of regulated financial documents: a case study on the auditor's report and memorandum of a French company. EDPACS, 1–9 (2025). https://doi.org/10.1080/07366981.2025.2549194
Bouziane, K., Echchaoui, M.: Leveraging artificial intelligence for international accessibility to Moroccan linguistic heritage: the case of translating Doukkala's ancient proverbs. Int. Arab J. Engl. Specif. Purp. **7**(1), 143–155 (2024). https://doi.org/10.34874/PRSM.iajesp-vol7iss1.48417
Briva-Iglesias, V., Dogru, G., Cavalheiro Camargo, J.L.: Large language models "ad referendum": how good are they at machine translation in the legal domain? MonTi Monografías De Traducción e Interpretación **16**, 75–107 (2024a). https://doi.org/10.6035/MonTI.2024.16.02
Briva-Iglesias, V., Camargo, J.L.C., Dogru, G.: Large language models "ad referendum": how good are they at machine translation in the legal domain? arXiv:2402.07681 (2024b). https://arxiv.org/abs/2402.07681
Elajail, I.B.: Evaluating the accuracy of machine translation in rendering Arabic idioms into English. North Afr. J. Sci. Publ. **3**(4), 36–48 (2025)
El-Saadany, M.: Artificial intelligence meets human expertise: a case study in translating violence against women reports. Res. Lang. Teach. **6**(31), 349–404 (2025)
Khan, M.I.: Impact of AI on legal and institutional translation: issues and strategies. Int. J. Transl. Interpretation Stud. **5**(6), 1–15 (2025). https://doi.org/10.32996/ijtis.2025.5.6.1
Khoshnevis, S., Alinouri, L.: AI versus Matecat in legal translation: accuracy and consistency in focus. Iran. J. Transl. Stud. **23**(90) (2025)
Moneus, A.M., Sahari, Y.: Artificial intelligence and human translation: a contrastive study based on legal texts. Heliyon **10**(6), e28106 (2024). https://doi.org/10.1016/j.heliyon.2024.e28106

Impact of Cultural Dimensions on the Acceptance of Artificial Intelligence in Moroccan Public Institutions: Analysis Based on Hofstede's Model

Abdelmounim Bouziane and Karima Bouziane

Abstract This study investigates how Hofstede's cultural dimensions and Technology Acceptance Model (TAM) factors predict AI adoption intentions among public sector employees in Casablanca, Morocco. Using a sample of 1424 respondents from key establishments, multiple linear regression explained 90.2% of variance in intention to use AI. Masculinity ($\beta = 0.765$) and collectivism ($\beta = 0.654$) emerged as strong positive predictors, reflecting Morocco's performance-driven, group-oriented culture that favors AI for collective efficiency gains. Conversely, uncertainty avoidance ($\beta = -0.645$) and power distance ($\beta = -0.534$) acted as significant barriers, underscoring bureaucratic rigidity and preference for hierarchical directives over untested innovations. These findings challenge TAM's universality, highlighting cultural moderators' primacy in high-context settings, and offer actionable strategies like leadership-endorsed pilots and team-based training to accelerate Morocco's Digital 2030 goals.

Keywords Artificial intelligence · Acceptance · Cultural dimensions · TAM theory · Public institutions · Morocco

1 Introduction

Artificial Intelligence (AI) represents a transformative force in public administration, promising enhanced efficiency, data-driven decision-making, and citizen-centric services. From automated permit processing to predictive resource allocation, AI could revolutionize Morocco's bureaucratic landscape—particularly in Casablanca,

A. Bouziane (✉)
Chouaib Doukkali University & Urban Agency of Casablanca, El Jadida, Morocco
e-mail: bouziane.a@ucd.ac.ma

K. Bouziane
ALCS Lab, Faculty of Letters and Humanities, Chouaib Doukkali University, El Jadida, Morocco

A. Bouziane and K. Bouziane (eds.), *AI, Management, and Culture: Shaping Organizations in the Digital Age*, Studies in Computational Intelligence 1291,
https://doi.org/10.1007/978-3-032-27882-1_5

the nation's financial capital hosting over 10,000 public sector employees across key establishments. Yet, technology adoption rarely follows a linear path. National culture profoundly shapes perceptions of innovation, as evidenced by Hofstede's framework where dimensions like uncertainty avoidance, power distance, and collectivism determine technology acceptance (Hofstede 1980; 2025 literature review). In Morocco—a high-context, hierarchical society with collectivist tendencies—these cultural factors likely interact uniquely with Technology Acceptance Model (TAM) constructs (perceived usefulness, ease of use) to influence AI adoption.

Research Gap: While Moroccan studies examine AI in education (Bouziane and Bouziane 2025), healthcare (Belharar et al. 2025), and SMEs (Erraoui and Amine 2024), no quantitative research integrates Hofstede's cultural dimensions with TAM to predict AI intention among public sector employees. Barodi and Lalaoui (2025) note 40% of civil servants cite skill gaps, yet cultural barriers remain unquantified.

Research Context: Casablanca's public establishments (Centre Régional d'Investissement, Agence Urbaine, Commune de Casablanca) serve as Morocco's administrative nerve center, employing 10,449 personnel. Their AI readiness directly impacts national digital transformation goals (Digital Morocco 2030).

Research Question: How do Hofstede's cultural dimensions and TAM factors predict AI adoption intention among Casablanca public sector employees?

2 Literature Review

National culture influences multiple domains of individuals' social, societal, and organizational life. Shared values, norms, and beliefs shape behaviors and communication patterns, relationships with authority, and the ways individuals perceive change, uncertainty, and decision-making processes.

In the fields of management and information systems, a substantial body of research has examined how cultural dimensions affect technology acceptance and use within organizations. Perceptions of usefulness, ease of use, trust, risk, and legitimacy may differ across national contexts. Consequently, the same technology (e.g., ERP systems, AI tools, collaborative platforms, or digital accounting processes) may be rapidly adopted in some cultures, while generating stronger resistance, caution, or a greater need for formal guidance in others.

This stream of literature therefore highlights the relevance of integrating national culture as an explanatory variable in technology acceptance models and digital transformation strategies, in order to better account for differences between usage intention, actual adoption, and effective appropriation of digital tools.

Research shows that national culture shapes the ethical, social, and practical acceptance of AI. Parrish and Saeed (2025) found that individuals from high power distance and masculine cultures are more likely to positively evaluate delegation of decision-making to AI-enabled systems, while uncertainty avoidance and indulgence predict more cautious attitudes. Jan et al. (2024) reported that individualism and uncertainty avoidance strongly influence perceived usefulness, ease of use, and behavioral intention in technology acceptance models (TAM, UTAUT). Similarly, O'Keefe and Minarik (2024) demonstrate that high uncertainty avoidance cultures exhibit lower trust and interdependence with AI teammates, suggesting that risk-averse employees may resist AI integration.

Barnes et al. (2024) emphasize that cultural identity affects AI perception: individuals in collectivist societies tend to perceive AI as an extension of the self, while those in individualistic societies may view it as a threat to autonomy and privacy. Sanguineti and Maran (2024) also observed that uncertainty avoidance reduces technology use, though perceived usefulness can mitigate these effects, highlighting the moderating role of context-specific factors.

AI acceptance is also shaped by broader socio-economic and organizational factors. Reiter et al. (2026) found that workers' economic expectations of AI vary across countries depending on socio-economic development and cultural tightness, with high human development levels sometimes associated with negative perceptions of AI. Murire (2024) highlights that AI adoption transforms organizational culture by promoting agility, continuous learning, and innovation, while also generating resistance due to fears of job displacement, ethical concerns, and skill shortages.

Trust is a recurring theme. Dang and Li (2025) identified AI capability, explainability, and cultural context as critical antecedents of trust, which in turn shapes behavioral intention. Chi et al. (2023) showed that trust and cultural dimensions moderate acceptance of AI service robots, emphasizing that AI adoption is not culturally neutral.

Studies in African and Middle Eastern countries underscore the relevance of cultural alignment in AI adoption. Shinkafi et al. (2025) report that collectivism, high power distance, and high uncertainty avoidance influence Intelligent Personal Assistant (IPA) adoption in African contexts. Alqaraleh et al. (2025) found that in collectivist societies, AI performance is evaluated primarily for group productivity rather than individual achievements, highlighting the importance of aligning AI implementation with local cultural norms.

Iamandi et al. (2024) showed that national culture affects the development of AI skills, with individualism negatively associated with AI talent concentration, while long-term orientation positively correlates with AI skill distribution. These findings suggest that cultural values influence not only adoption but also the ecosystem of AI competencies within organizations.

In the Moroccan context, studies examining how cultural dimensions shape the acceptance of artificial intelligence (AI) remain relatively limited and fragmented. Existing work tends to focus on a small set of fields, mainly education (especially debates around the use and regulation of generative AI), healthcare, and finance/audit-related practices.

This sectoral focus indicates that many Moroccan organizational settings are still under-researched (e.g., industry, public administration, HR, SMEs), even though AI acceptance can be influenced by sociocultural factors such as trust, perceived fairness, transparency expectations, and uncertainty tolerance.

Bouziane and Bouziane (2025) found that Moroccan university students' attitudes toward AI vary by institutional affiliation, with private universities showing higher enthusiasm. Darban et al. (2024) identified trust, technological infrastructure, and cultural factors as key determinants of AI adoption among Moroccan consumers.

In SMEs, Erraoui and Amine (2024) observed that perceived ease of use, usefulness, adoption barriers, and environmental factors interact with organizational culture to shape AI adoption decisions. In healthcare, Belharar et al. (2025) highlighted that AI adoption is facilitated by skills development, infrastructural improvement, and culturally adapted tools, while fears of dehumanization and ethical concerns hinder acceptance.

In the public sector, Barodi and Lalaoui (2025) reported that younger civil servants are more receptive to AI, but over 40% identified insufficient digital skills as a barrier, highlighting the need for effective change management. Benhayoun et al. (2025) found that in auditing, ease of use significantly predicts AI readiness, whereas top management commitment and perceived usefulness are less influential. Ouzif et al. (2025) showed that facilitating conditions and social influence are key determinants of AI adoption among university professors. Hanaty (2025) identified leadership engagement and a culture of innovation as decisive factors for AI effectiveness in Moroccan institutions.

Finally, Jouiet and El Ghaleb (2025) demonstrated that technological, organizational, and environmental barriers—such as infrastructure limitations, regulatory gaps, and lack of technical assistance—impede AI integration in public sector internal audit functions. Collectively, these studies indicate that Moroccan AI adoption is shaped by cultural values, organizational support, and individual readiness, with uncertainty avoidance, collectivism, and institutional norms emerging as key factors.

The reviewed literature highlights that AI adoption is a socio-technical process influenced by both cultural and organizational factors. Hofstede's dimensions—particularly individualism, uncertainty avoidance, and masculinity—emerge as critical predictors of AI acceptance. Across African, Middle Eastern, and Moroccan contexts, cultural values moderate perceptions of AI's usefulness, trustworthiness, and social appropriateness.

Despite these insights, there is a gap in quantitative research linking cultural dimensions to AI adoption in Moroccan public institutions, particularly studies that integrate Hofstede's framework with technology acceptance variables. Addressing this gap is crucial for developing culturally aligned AI deployment strategies that can enhance the success of digital transformation initiatives in Morocco.

3 Methodology

This study employs a quantitative approach to examine the acceptance of artificial intelligence (AI) among employees in public establishments in Casablanca, Morocco's financial capital and a key economic hub. Casablanca was selected due to its concentration of public institutions and its strategic importance to the national economy.

Table 1 Sampling details

Establishment	Total personnel	Responses received
Centre Régional d'Investissement	50	22
Agence Urbaine de Casablanca	199	82
Commune de Casablanca	10,000	1320
Société Régionale Multiservices	200	122
Total	10,449	1424

3.1 Sample Composition

A purposive sample was drawn from four major public establishments in Casablanca, targeting a diverse range of organizational sizes and functions. Questionnaires were distributed electronically and in person to a total population of 10,449 employees. A total of 1424 valid responses were received, yielding an overall response rate of approximately 13.6%. The breakdown is as follows (Table 1).

3.2 Data Collection Instrument

Data were collected via a structured questionnaire administered between February 2025 and June 2025. The instrument comprised Likert-scale items adapted from Hofstede's cultural dimensions framework, integrated with the Technology Acceptance Model (TAM) to assess AI adoption. Key dimensions and sample items include (Table 2):

The questionnaire was pre-tested on a pilot sample of 50 employees (Cronbach's $\alpha > 0.80$ for all scales), ensuring reliability and cultural relevance through back-translation from English to French and Arabic.

3.3 Data Analysis

Responses were analyzed using IBM SPSS Statistics (version [insert version, e.g., 27]). Preliminary analyses included descriptive statistics (means, standard deviations) and reliability tests (Cronbach's α). The primary analysis involved multiple linear regression to identify predictors of AI intention to use, with cultural dimensions, perceived usefulness, ease of use, trust, and perceived risk as independent variables. Assumptions (linearity, multicollinearity via VIF < 5, normality of residuals) were verified. The model equation is:

Table 2 Research variables—questionnaire

Individualism versus collectivism
I prefer working in a team rather than individually to achieve common goals
Loyalty to my group/organization takes precedence over my personal ambitions
A project's success depends more on collective effort than individual initiatives
Uncertainty avoidance
New technologies like AI make me anxious if their functioning is not perfectly clear
I prefer strict rules and well-defined procedures before adopting AI
Technological innovations should be extensively tested before deployment
Masculinity versus femininity
Performance and concrete results are more important than harmonious workplace relationships
Competition and personal achievement motivate me more than benevolent collaboration
My organization values challenges and risks over quality of life
Power distance
Decisions about AI adoption should come from top management, not employees
I respect authority and expect clear instructions before using AI
I use AI without having instructions from top management
Perceived usefulness
AI would significantly improve my efficiency in daily tasks
With AI, I could accomplish more work in less time
AI will have no significant impact on my tasks
Ease of use
AI is easy to use, even without extensive training
Interacting with AI requires minimal mental effort
I need training to use AI
Trust and perceived risk
I trust AI to make objective and reliable decisions
Using AI involves too many risks (errors, biases, privacy issues)
Intention to use (dependent variable)
I am willing to use AI in my work if it becomes available
I prefer not to use AI in my work

$$\begin{aligned}\text{Intention to Use} = {} & \beta 0 + \beta 1(\text{Cultural Dimensions}) + \beta 2(\text{Perceived Usefulness}) \\ & + \beta 3(\text{Ease of Use}) + \beta 4(\text{Trust}) + \beta 5(\text{Perceived Risk}) + \epsilon\end{aligned}$$

Ethical considerations included informed consent, anonymity, and approval from institutional review boards where required.

Table 3 Statistical tests

Statistic	Value	Interpretation
R^2 (adjusted)	0.902	The model explains 90.2% of the variance in intention to use AI, indicating excellent explanatory power
F-statistic (*p*-value)	0.000	Highly significant ($p < 0.001$), rejecting the null hypothesis that all coefficients equal zero
Durbin-Watson (DW)	1.98	Close to 2.0, confirming no significant autocorrelation in residuals (ideal range: 1.5–2.5)

4 Results

Prior to extracting correlation coefficients, diagnostic tests confirmed the validity of the multiple linear regression model. The key statistics are summarized below (Table 3).

These results validate the assumptions of linearity, homoscedasticity, and independence, justifying the use of multiple linear regression.

The standardized beta coefficients (β) from the regression model are presented below, with the dependent variable being Intention to Use AI. Positive coefficients indicate a positive relationship with AI adoption intent; negative coefficients indicate barriers (Table 4).

Model Equation:

$$\begin{aligned}\text{Intention to Use} = {} & 0.765(\text{Masculinity}) + 0.654(\text{Collectivism}) \\ & - 0.645(\text{Uncertainty Avoidance}) - 0.534(\text{Power Distance}) \\ & - 0.501(\text{Usefulness}) - 0.403(\text{Ease of Use}) + 0.003(\text{Trust}) + \epsilon\end{aligned}$$

The results align with Morocco's public sector characteristics. High Masculinity and Collectivism ($\beta = 0.765, 0.654$) reflect a performance-driven yet group-oriented workforce receptive to AI for collective gains. However, Uncertainty Avoidance and Power Distance ($\beta = -0.645, -0.534$) mirror bureaucratic rigidity, older civil servants' resistance to change, and centralized decision-making prevalent in Casablanca's administrations.

5 Discussion

The findings reveal that cultural dimensions and technology acceptance factors significantly predict AI adoption intentions among Casablanca public sector employees, explaining 90.2% of variance (R2 = 0.902). These results both confirm and challenge existing literature, offering novel insights into Morocco's bureaucratic context.

Table 4 Linear regression coefficients

Predictor	β coefficient	Significance level	Interpretation
Individualism versus collectivism	0.654	$p < 0.01$	Strongest positive predictor: collectivist orientations enhance AI adoption, likely due to Morocco's high-context culture favoring group consensus on technological integration
Masculinity versus femininity	0.765	$p < 0.001$	Most influential factor: achievement-oriented, competitive cultures strongly drive AI acceptance, prioritizing performance gains over relational harmony
Uncertainty avoidance	– 0.645	$p < 0.01$	Major barrier: high uncertainty avoidance (common in public administration) generates anxiety toward opaque AI systems, preferring tested procedures
Power distance	– 0.534	$p < 0.05$	Significant inhibitor: hierarchical structures delay AI adoption, with employees awaiting top-down directives rather than proactive use
Perceived usefulness	– 0.501	$p < 0.05$	Counterintuitive barrier: despite recognizing efficiency gains, perceived risks may outweigh benefits in bureaucratic contexts
Ease of use	– 0.403	$p < 0.05$	Moderate barrier: concerns about training needs hinder adoption, particularly among older staff
Trust and perceived risk	0.003	$p > 0.05$	Non-significant: neutral impact, suggesting trust issues balance with risk perceptions
Intention to use (DV)	1.000	–	Dependent variable (standardized)

The strongest predictor, Masculinity versus Femininity (β = 0.765), aligns with Parrish and Saeed (2025), who found masculine cultures favor AI delegation for performance gains. Moroccan public employees' achievement orientation mirrors this, prioritizing "concrete results" over relational harmony—a pragmatic response to administrative pressures. Similarly, Individualism versus Collectivism (β = 0.654) supports Barnes et al. (2024) and Shinkafi et al. (2025), as collectivist tendencies enhance AI acceptance when framed as group productivity tools rather than individual threats.

High Uncertainty Avoidance (β = – 0.645) and Power Distance (β = – 0.534) emerge as major inhibitors, corroborating Jan et al. (2024), O'Keefe and Minarik (2024), and Alqaraleh et al. (2025). These reflect Morocco's bureaucratic DNA—rigid procedures, hierarchical decision-making, and anxiety toward "untested" AI—echoed in Barodi and Lalaoui (2025), where 40% of civil servants cited digital skill

Table 5 Comparison with Moroccan studies

Dimension/factor	This study (β)	Moroccan literature	Alignment
Masculinity	+ 0.765	N/A	Novel finding
Collectivism	+ 0.654	Shinkafi et al. (2025)	Strong
Uncertainty avoidance	– 0.645	Darban et al. (2024)	Strong
Power distance	– 0.534	Jouiet and El Ghaleb (2025)	Strong
Perceived usefulness	– 0.501	Benhayoun et al. (2025)	Partial (contradicts)

gaps. The preference for "strict rules before adopting AI" and top-down directives directly matches Benhayoun et al. (2025).

The counterintuitive negative Perceived Usefulness ($\beta = -0.501$) challenges core TAM assumptions (Davis 1989) and Erraoui and Amine (2024). Despite recognizing efficiency gains, risk perceptions dominate in high-stakes public administration, supporting Reiter et al. (2026) on socio-economic caution. Ease of Use concerns ($\beta = -0.403$) align with Barodi and Lalaoui (2025), while non-significant Trust/ Perceived Risk ($\beta = 0.003$) suggests balanced ambivalence, differing from Dang and Li (2025).

Age-related resistance, bureaucratic inertia, and administrative rigidity—prevalent among Casablanca's 10,000+ municipal workers—exacerbate cultural barriers, explaining why younger civil servants show greater receptivity (Barodi and Lalaoui 2025).

Comparison with Moroccan Studies

The comparison with previous studies conducted within the Moroccan context can be as follow (Table 5).

This study fills the quantitative gap identified in the literature review, being the first to integrate Hofstede's framework with TAM in Moroccan public administration using a large sample ($n = 1424$).

6 Recommendations

The regression results provide actionable insights for enhancing AI adoption in Casablanca's public establishments. Recommendations are prioritized by impact (β coefficients) and organized by stakeholder.

1. *Leadership and Policy Interventions (Addressing Power Distance: $\beta = -0.534$)*

    - Top-down endorsement: Public institution directors should issue formal AI adoption policies with clear directives, reducing employee wait for hierarchical approval.

- Pilot mandates: Require 20% of departments to implement AI tools (e.g., chatbots for citizen queries) within 6 months, demonstrating leadership commitment.
- National framework: Ministry of Digital Transition to develop AI guidelines for public sector, mirroring Singapore's GovTech model adapted to Moroccan hierarchy.

2. *Cultural Alignment Strategies (Leveraging Masculinity: $\beta = 0.765$ & Collectivism: $\beta = 0.654$)*

- Performance framing: Market AI as "efficiency champions" with KPI dashboards showing time savings (e.g., "Reduce report processing from 3 days to 3 h").
- Team-based rollout: Deploy AI through cross-departmental "AI task forces" emphasizing collective gains over individual use.
- Competitive incentives: Launch "AI Innovation Awards" for departments achieving highest adoption rates, appealing to achievement-oriented staff.

3. *Training and Uncertainty Reduction (Addressing Uncertainty Avoidance: $\beta = -0.645$)*

- Hands-on workshops: 2-day practical training focusing on "AI explainability" (how algorithms work), targeting the 1320 Commune de Casablanca respondents.
- Progressive testing: Implement 3-month "sandbox" phases where AI handles low-risk tasks (document classification) before scaling.
- Moroccan case studies: Develop training materials featuring successful AI implementations in OCP or Maroc Telecom to build local legitimacy.

4. *Targeted Demographics and Change Management (Addressing Ease of Use: B = − 0.403 & Age Resistance)*

Age-segmented programs:

The strategies to make to address the ease of use and the age of resistance can be structured as follows (Table 6).

Change champions: Identify 10% of early adopters (high masculinity scorers) as peer trainers for resistant colleagues.

5. *Implementation Roadmap for Casablanca Establishments*

The roadmap to implement to boast the use of AI in public institutions can be as follows (Table 7).

Table 6 Age-segmented programs

Age group	Strategy	Priority
18–35	Self-paced digital modules	High
36–50	Hybrid workshops + mentors	Medium
50+	1:1 coaching + simplified interfaces	High

Table 7 Roadmap for Casablanca Establishments

Establishment	Tailored strategy	Expected impact
Centre Régional d'Investissement (n = 22)	Quick-win chatbots for investor queries	80% adoption in 3 months
Agence Urbaine (n = 82)	Document automation pilots	Reduce processing by 40%
Commune de Casablanca (n = 1320)	Citizen service AI (complaints, permits)	Handle 30% of routine queries

These recommendations transform cultural barriers into adoption accelerators, positioning Casablanca's public sector as Morocco's AI vanguard while respecting national values and bureaucratic realities.

7 Conclusion

This study provides robust evidence (R2 = 0.902, F < 0.001) that cultural dimensions dominate AI adoption intentions among Casablanca's public sector employees. Masculinity (β = 0.765) and Collectivism (β = 0.654) emerge as powerful accelerators, confirming Morocco's performance-driven yet group-oriented workforce responds positively when AI delivers collective gains. Conversely, Uncertainty Avoidance (β = − 0.645) and Power Distance (β = − 0.534) constitute formidable barriers, reflecting bureaucratic realities where employees await hierarchical directives and proven procedures.

These findings advance theory by challenging TAM's universalism, demonstrating cultural moderators' primacy in high-context settings. Masculinity surpasses traditional predictors (usefulness, trust), while counterintuitive negative Perceived Usefulness (β = − 0.501) reveals risk perceptions trumping efficiency gains in high-stakes administration.

Key Contributions

- First quantification: Hofstede-TAM integration in Moroccan public sector (n = 1424).
- Contextual amplifiers: Age, bureaucracy, and administrative rigidity explain effect sizes.
- Policy roadmap: Leadership endorsement, team-based rollouts, and progressive training transform barriers into accelerators.

Limitations

- Cross-sectional design: Captures intentions, not actual adoption. Longitudinal tracking needed.
- Self-reported data: Potential social desirability bias in hierarchical context.

- Sample representation: 13.6% response rate; non-respondents may differ systematically.
- Cultural dimensions: Limited to four Hofstede factors; Long-Term Orientation/ Indulgence excluded.
- Context specificity: Casablanca-centric; national/generalizability requires validation.

Future research should prioritize longitudinal studies tracking actual AI adoption behaviors beyond self-reported intentions, establishing causality and measuring tangible ROI in Casablanca's public establishments. Comparative analyses between public and private sectors would reveal if cultural barriers persist across organizational types, while qualitative investigations into non-respondents (86.4% of the population) could uncover hidden resistance patterns among older civil servants exhibiting high uncertainty avoidance. Expanding the model to include all six Hofstede dimensions—particularly Long-Term Orientation and Indulgence—alongside UTAUT2 constructs would provide a comprehensive framework, validated through multi-regional samples (Rabat, Marrakech, Tangier) for national generalizability. Emerging priorities include sector-specific applications like generative AI (ChatGPT) in citizen services and experimental designs testing intervention effectiveness, such as leadership endorsement programs or gamified training. Cross-national comparisons with other high-context African/Middle Eastern countries would further contextualize Morocco's unique masculinity-collectivism profile, informing pan-regional digital transformation strategies.

References

Alqaraleh, M.K.S., Hannoon, A., Mahmood, A.: Emerging technologies in the Middle East: artificial intelligence adoption and performance expectancy in SMEs. In: Hannoon, A., Mahmood, A. (eds.) Intelligence-Driven Circular Economy, pp. 215–235. Springer (2025). https://doi.org/10.1007/978-3-031-73899-9_15

Barnes, A.J., Zhang, Y., Valenzuela, A.: AI and culture: culturally dependent responses to AI systems. Curr. Opin. Psychol. **58**, 101838 (2024). https://doi.org/10.1016/j.copsyc.2024.101838

Barodi, M., Lalaoui, S.: Civil servants' readiness for AI adoption: the role of change management in Morocco's public sector. Prob. Perspect. Manag. **23**(1), 63–75 (2025). https://doi.org/10.21511/ppm.23(1).2025.05[2]

Belharar, O., Amrani, S., Chakor, A.: Facilitators and barriers to the adoption of artificial intelligence by Moroccan nurses: an exploratory qualitative study. Policies, Adm. Markets J. **1**(1–2) (2025)

Benhayoun, I., Bougrine, S., Sassioui, A.: Readiness for artificial intelligence adoption by auditors in emerging countries – a PLS-SEM analysis of Moroccan firms. J. Financ. Report. Account. **23**(4), 1486–1508 (2025). https://doi.org/10.1108/JFRA-07-2024-0448

Bouziane, A., Bouziane, K.: Analysis of artificial intelligence acceptance in humanities and social sciences: the case of Moroccan universities. Discov. Educ. **4**, 424 (2025). https://doi.org/10.1007/s44217-025-00632-1

Chi, O.H., Chi, C.G., Gursoy, D., Nunkoo, R.: Customers' acceptance of artificially intelligent service robots: the influence of trust and culture. Int. J. Inf. Manage. **70**, 102623 (2023). https://doi.org/10.1016/j.ijinfomgt.2023.102623

Dang, Q., Li, G.: Unveiling trust in AI: the interplay of antecedents, consequences, and cultural dynamics. AI & Soc. (2025). https://doi.org/10.1007/s00146-025-02477-6

Darban, K., Kabbaj, S., Eljai, M.: Assessing the adoption readiness of Moroccan consumers for AI-powered assistance and CRM systems. Procedia Comput. Sci. **236**, 541–549 (2024). https://doi.org/10.1016/j.procs.2024.05.064

Davis, F.: A technology acceptance model for empirically testing new end-user information systems: theory and results. Doctoral dissertation, MIT Sloan School of Management, Cambridge, MA (USA) (1986)

Erraoui, S., Amine, A.: Proposal of a technology acceptance model: adoption of artificial intelligence in Moroccan SMEs. Eur. J. Econ. Financ. Res. **8**(6), 79 (2024). https://doi.org/10.46827/ejefr.v8i6.1858

Hanaty, H.: Determinants of artificial intelligence effectiveness in Moroccan public institutions and enterprises. Sci. Res. Innov. Stud. **4**(2) (2025)

Iamandi, I.E., Constantin, L.G., Munteanu, S.M., Cernat-Gruici, B.: Insights on the relationship between artificial intelligence skills and national culture. Amfiteatru Econ. **26**(67), 741–761 (2024). https://doi.org/10.24818/EA/2024/67/741

Jan, J., Alshare, K.A., Lane, P.L.: Hofstede's cultural dimensions in technology acceptance models: a meta-analysis. Univ. Access Inf. Soc. **23**, 717–741 (2024). https://doi.org/10.1007/s10209-022-00930-7

Jouiet, I., El Ghaleb, N.E.A.: Barriers to the integration of artificial intelligence in public sector internal audit in Morocco: an exploratory study. Edelweiss Applied Science and Technology **9**(5), 1291–1302 (2025)

Murire, O.T.: Artificial intelligence and its role in shaping organizational work practices and culture. Adm. Sci. **14**(12), 316 (2024). https://doi.org/10.3390/admsci14120316

O'Keefe, G., Minarik, M.: Culture and AI: The impact of uncertainty avoidance on human–AI teammate interdependence and trust. Faculty presentation, October 31. Molloy University, DigitalCommons@Molloy (2024). https://digitalcommons.molloy.edu/. Last accessed 2025/12/29

Ouzif, H., El Boukhari, H., El Achabi, M., Chahbouni, O., Essanoun, M., Mahouat, N., Bakkali, S.: Analyzing the adoption of artificial intelligence by Moroccan university teachers: Key insights and implications from the UTAUT model. Edelweiss Appl. Sci. Technol. **9**(4), 2722–2732 (2025). https://doi.org/10.55214/25768484.v9i4.6644

Parrish, J., Saeed, K.: Understanding the impact of culture on the teleological evaluation of delegation to artificial intelligence-enabled information systems. Technol. Forecast. Soc. Chang. **219**, 124247 (2025). https://doi.org/10.1016/j.techfore.2025.124247

Reiter, L., Böhm, R., Fuchs, C.: Country-level differences in socio-economic development and cultural dimensions are associated with workers' economic expectations of AI: evidence from 31 countries. Comput. Hum. Behav. Rep. **21**, 100905 (2026). https://doi.org/10.1016/j.chbr.2025.100905

Sanguineti, F., Maran, T.: Navigating the clouds: cultural influences on technology acceptance in the workplace. Technol. Soc. **77**, 102553 (2024). https://doi.org/10.1016/j.techsoc.2024.102553

Shinkafi, A.A., Bassey, S., Chaku, S.E., Aimufua, G.I.O., Joseph, A.D.: Cultural dimensions and their impact on intelligent personal assistant adoption in African contexts. Afr. Multi. J. Sci. Artif. Intell. **2**(2), 392–405 (2025). https://doi.org/10.58578/amjsai.v2i2.6481

Impact of University Life Dynamics on the Organizational Climate of Companies in the Era of Artificial Intelligence

Samah Fifani

Abstract This article examines the influence of university life dynamics on the organizational climate of companies integrating artificial intelligence (AI) into management and production practices. The study analyzes how experiences at university, including participation in sports, associations, entrepreneurial or volunteer activities, shape employee behaviors, values, and adaptability to new technologies. The methodology relies on a quantitative survey of 450 employees from various sectors in Morocco, using validated psychometric scales and statistical analyses (SPSS 27). Results show that alumni involved in sports and volunteering positively contribute to the organizational climate ($\beta = 0.487$, $p < 0.001$), fostering cooperation, teamwork, and resilience to technological changes. In contrast, those focused on entrepreneurship or with limited extracurricular experience have a more neutral or sometimes negative impact ($\beta = -0.156$, $p < 0.05$), often due to integration difficulties or individualistic approaches. This research highlights the importance of social and behavioral skills acquired during university as essential leverage for adapting to organizational contexts in the AI era.

Keywords University life · Organizational climate · Artificial intelligence · Social skills—employability · Technological adaptation · Morocco

1 Introduction

The integration of artificial intelligence (AI) into organizational processes represents a major transformation that redefines work dynamics, required skills, and organizational climate (Brynjolfsson and McAfee 2014; Dwivedi et al. 2021). In this context of

S. Fifani (✉)
Chouaib Doukkali University, El Jadida, Morocco
e-mail: fifani.samah@ucd.ac.ma

A. Bouziane and K. Bouziane (eds.), *AI, Management, and Culture: Shaping Organizations in the Digital Age*, Studies in Computational Intelligence 1291,
https://doi.org/10.1007/978-3-032-27882-1_6

rapid technological evolution, human and social competencies become critical differentiators for organizational success (Chui et al. 2018). Moroccan companies, increasingly adopting AI technologies, face challenges related to employee adaptation and maintaining a positive organizational climate (Fifani 2026).

University education, beyond academic knowledge transmission, constitutes a crucial space for developing transversal competencies through extracurricular activities (Pascarella and Terenzini 2005; Kuh 2008). Sports participation, associative engagement, volunteering, and entrepreneurial initiatives offer opportunities to develop leadership, teamwork, resilience, and adaptability skills (Holt et al. 2017; Rubin et al. 2014).

Despite the recognized importance of these experiences, few empirical studies have examined their specific impact on organizational climate in companies integrating AI, particularly in the Moroccan context. This research addresses this gap by investigating: **How do university life dynamics influence the organizational climate of Moroccan companies adopting artificial intelligence technologies?**

Research objectives

1. Identify the types of university activities influencing organizational climate
2. Measure the impact of different extracurricular experiences on employee adaptation to AI
3. Analyze the mediating role of social and behavioral skills
4. Provide recommendations for universities and companies

2 Literature Review and Theoretical Framework

2.1 Organizational Climate in the AI Era

Organizational climate refers to employees' shared perceptions of organizational policies, practices, and procedures (Schneider et al. 2013). In AI-integrated environments, this climate is affected by technological uncertainty, job transformation fears, and new collaboration requirements between humans and machines (Jarrahi 2018; Wilson and Daugherty 2018).

Recent studies indicate that successful AI adoption depends on factors including trust, communication, and organizational support (Davenport and Ronanki 2018). Naranjo-Valencia et al. (2016) demonstrate that innovative organizational climates facilitate technology adoption, while Brock and Von Wangenheim (2019) emphasize the importance of human skills in AI contexts.

2.2 University Life and Competency Development

Extracurricular activities during university constitute significant learning spaces (Kuh 2009; Astin 1999). Research identifies several key dimensions:

Sports and Athletics Develop discipline, teamwork, stress management, and leadership (Fraser-Thomas et al. 2005; Eccles et al. 2003). Studies show positive correlations between sports participation and professional success (Kniffin et al. 2015).

Volunteering and Civic Engagement Foster empathy, social responsibility, prosocial behaviors, and cooperation capacity (Astin et al. 2000; Celio et al. 2011). These experiences enhance adaptability and interpersonal skills (Jones and Hill 2003).

Entrepreneurial Activities Develop initiative, risk-taking, innovation, and autonomy (Rauch and Hulsink 2015). However, they may also reinforce individualistic behaviors potentially conflicting with collaborative organizational climates (Hmieleski and Baron 2009).

Associative Participation Strengthens organizational, communication, and networking skills (Foubert and Grainger 2006). These experiences facilitate organizational integration and collective project engagement.

2.3 Conceptual Model and Hypotheses

Based on Social Learning Theory (Bandura 1977) and Human Capital Theory (Becker 1964), we propose the following model:

H1 Sports participation during university positively influences organizational climate perception in AI-integrated companies.

H2 Volunteering and civic engagement positively impact organizational climate and adaptation to AI technologies.

H3 Entrepreneurial activities have a complex (potentially negative) relationship with organizational climate due to individualistic tendencies.

H4 Limited extracurricular engagement negatively affects organizational climate contribution.

H5 Social and behavioral skills mediate the relationship between university activities and organizational climate.

3 Methodology

3.1 Research Design and Sample

This quantitative research employs a cross-sectional survey design. The target population comprises employees in Moroccan companies having integrated AI technologies within the last three years.

Sampling

- **Sampling method**: Stratified random sampling by sector and company size
- **Sample size**: 450 respondents (margin of error: 4.6%, confidence level: 95%)
- **Distribution**: Technology/IT (35%), Finance/Banking (28%), Manufacturing (22%), Services (15%)
- **Response rate**: 78.3%

Participant characteristics

- Age: 24–45 years (M = 32.4, SD = 5.8)
- Gender: 58% male, 42% female
- Education: 100% university graduates (Bachelor's: 42%, Master's: 48%, PhD: 10%)
- Professional experience: 3–15 years (M = 7.2, SD = 3.4)

3.2 Measurement Instruments

- **University activity engagement scale (UAES)**—Adapted from Kuh (2009)
  - Sports participation (5 items, α = 0.87)
  - Volunteering (5 items, α = 0.89)
  - Entrepreneurship (5 items, α = 0.84)
  - Associative participation (5 items, α = 0.86)
  - 5-point Likert scale (1 = Never, 5 = Very frequently)
- **Organizational climate scale (OCS)**—Patterson et al. (2005)
  - Cooperation and teamwork (6 items, α = 0.91)
  - Innovation orientation (5 items, α = 0.88)
  - Communication quality (5 items, α = 0.90)
  - 7-point Likert scale (1 = Strongly disagree, 7 = Strongly agree)
- **AI Adaptation scale (AIAS)**—Developed and validated for this study
  - Technological openness (4 items, α = 0.85)
  - Learning agility (4 items, α = 0.87)
  - Change resilience (4 items, α = 0.89)

  - 5-point Likert scale

- **Social and behavioral skills scale (SBSS)**—Adapted from OECD (2015)
  - Emotional intelligence (4 items, $\alpha = 0.88$)
  - Collaboration skills (4 items, $\alpha = 0.90$)
  - Leadership (4 items, $\alpha = 0.86$)
  - Adaptability (4 items, $\alpha = 0.87$)

3.3 Data Collection and Analysis

Data collection occurred between Septembre-Novembre 2025 via online questionnaires distributed through company HR departments. Participants provided informed consent, and anonymity was guaranteed.

Statistical analyses (SPSS 27.0)

- Descriptive statistics and frequency distributions
- Reliability analysis (Cronbach's alpha)
- Exploratory factor analysis (EFA) with KMO and Bartlett tests
- Correlation analysis (Pearson's r)
- Multiple regression analysis
- Mediation analysis (Hayes PROCESS macro, Model 4)
- ANOVA for group comparisons

4 Results

4.1 Reliability and Validity Assessment

All scales demonstrate excellent internal consistency ($\alpha > 0.80$) and sampling adequacy (KMO > 0.79), with significant Bartlett tests confirming factorability (Table 1).

4.2 Descriptive Statistics

Strong positive correlations exist between sports, volunteering, and organizational climate outcomes. Entrepreneurship shows weak or negative associations (Table 2).

Table 1 Reliability statistics

Scale	Items	Cronbach's α	KMO	Bartlett's χ^2	*p*-value
Sports participation	5	0.872	0.841	892.45	<0.001
Volunteering	5	0.894	0.867	1024.67	<0.001
Entrepreneurship	5	0.841	0.798	745.23	<0.001
Organizational climate	16	0.935	0.924	2847.91	<0.001
AI adaptation	12	0.911	0.889	1876.34	<0.001
Social skills	16	0.927	0.912	2634.78	<0.001

Source SPSS Output

Table 2 Descriptive statistics and correlations

Variable	M	SD	1	2	3	4	5	6
1. Sports	3.24	1.12	–					
2. Volunteering	2.87	1.24	.456**	–				
3. Entrepreneurship	2.45	1.18	.234**	.189**	–			
4. Org. Climate	5.12	0.94	.512**	.498**	−.078	–		
5. AI adaptation	3.89	0.87	.467**	.421**	.156*	.623**	–	
6. Social skills	4.21	0.79	.541**	.523**	.098	.687**	.594**	–

Source SPSS Output
Note **$p < .01$, *$p < .05$, N = 450

4.3 Group Differences Analysis

Post-hoc Tukey tests reveal significant differences ($p < 0.001$) between high sports/volunteering groups versus entrepreneurship/low engagement groups, supporting H1, H2, and H4 (Table 3).

Table 3 ANOVA Results—Organizational climate by activity type

Activity Profile	N	M	SD	F	P	η^2
High sports	124	5.67	0.71	34.56	<0.001	0.189
High volunteering	98	5.54	0.78			
High entrepreneurship	87	4.82	1.02			
Low engagement	141	4.67	0.95			

Source SPSS Output

Table 4 Multiple regression—Predictors of organizational climate

Predictor	B	SE	β	t	P	95% CI
(Constant)	2.145	0.234	–	9.167	<0.001	[1.685, 2.605]
Sports participation	0.408	0.042	.487**	9.714	<0.001	[0.326, 0.490]
Volunteering	0.356	0.038	.471**	9.368	<0.001	[0.281, 0.431]
Entrepreneurship	−0.124	0.039	−.156*	−3.179	.002	[−0.201, −0.047]
Low engagement	−0.287	0.045	−.289**	−6.378	<0.001	[−0.375, −0.199]
Age	0.012	0.008	.067	1.500	.134	[−0.004, 0.028]
Gender	−0.045	0.089	−.024	−0.506	.613	[−0.220, 0.130]

Source SPSS Output
Note: $R^2 = 0.456$, Adjusted $R^2 = 0.448$, $F(6,443) = 62.34$, $p < 0.001$

4.4 Regression Analysis

The model explains 45.6% of variance in organizational climate. Sports and volunteering are strong positive predictors, while entrepreneurship and low engagement negatively predict climate perceptions, fully supporting H1-H4 (Table 4).

4.5 Mediation Analysis

Social skills significantly mediate the relationship between university activities and organizational climate, with 57.4% of the sports effect transmitted through social skills development, supporting H5 (Table 5).

Table 5 Mediation effects of social skills

Path	Effect	SE	95% CI	Result
Sports → social skills → org. Climate	0.234	0.038	[0.162, 0.312]	Partial mediation
Volunteering → social skills → org. Climate	0.198	0.041	[0.119, 0.281]	Partial mediation
Direct effect (sports)	0.174	0.045	[0.086, 0.262]	Significant
Total effect (sports)	0.408	0.042	[0.326, 0.490]	Significant
Proportion mediated	57.4%	–	–	–

Source SPSS Output
Note: Bootstrap samples = 5000, Bias-corrected confidence intervals

Table 6 Regression—University activities predicting AI adaptation

Predictor	B	SE	β	t	P
Sports participation	0.342	0.048	.441**	7.125	<0.001
Volunteering	0.287	0.045	.409**	6.378	<0.001
Entrepreneurship	0.116	0.046	.157*	2.522	.012
Social skills	0.394	0.055	.358**	7.164	<0.001

Source SPSS Output
Note: $R^2 = 0.427$, $F(4,445) = 83.15$, $p < 0.001$

4.6 AI Adaptation Analysis

Sports and volunteering strongly predict AI adaptation, with social skills serving as a significant mechanism (Table 6).

5 Discussion

5.1 Principal Findings

This research demonstrates that university life dynamics significantly influence organizational climate in AI-integrated companies. Three major findings emerge:

- **Positive impact of sports and volunteering**: Results strongly support H1 and H2, confirming that sports ($\beta = 0.487$) and volunteering ($\beta = 0.471$) participants substantially contribute to positive organizational climates. This aligns with previous research showing these activities develop cooperation, resilience, and teamwork skills (Kniffin et al. 2015; Astin et al. 2000). In AI contexts requiring human-machine collaboration, these competencies prove essential for maintaining cohesion and managing technological transitions.
- **Ambivalent entrepreneurship effects**: H3 receives support through entrepreneurship's negative coefficient ($\beta = -0.156$), suggesting entrepreneurial mindsets may conflict with collaborative organizational requirements. This finding nuances existing literature generally praising entrepreneurial education (Rauch & Hulsink, 2015), indicating that individualistic orientations developed through entrepreneurship can hinder organizational integration, particularly in AI environments demanding intensive collaboration.
- **Critical role of extracurricular engagement**: H4 confirmation ($\beta = -0.289$) highlights risks associated with limited university involvement. Employees lacking extracurricular experience demonstrate lower organizational climate contributions and reduced AI adaptation capacity, emphasizing holistic education importance beyond pure academic achievement.

- **Social skills mediation**: H5 validation reveals 57.4% of university activities' effects transit through social skills, confirming these competencies as fundamental mechanisms linking university experiences to organizational outcomes. This finding supports calls for "human skills" in AI era (Wilson and Daugherty 2018; Chui et al. 2018).

5.2 Theoretical Contributions

This research advances theoretical understanding through several contributions:

First, it extends organizational climate theory to AI contexts, demonstrating climate's continued relevance amid technological transformation while identifying specific factors enabling positive climate maintenance.

Second, it bridges higher education and organizational behavior literature by empirically linking university experiences to workplace outcomes in technology-driven environments, addressing calls for this integration (Pascarella and Terenzini 2005).

Third, it provides nuanced understanding of entrepreneurship's organizational effects, suggesting context-dependent impacts requiring further investigation.

5.3 Practical Implications

For universities

- Strengthen extracurricular activity opportunities, particularly sports and volunteering programs
- Integrate social skills development into curricula
- Balance entrepreneurial education with collaborative competency development
- Create awareness among students regarding extracurricular participation's professional value

For companies

- Consider extracurricular backgrounds in recruitment processes as organizational climate predictors
- Develop onboarding programs compensating for limited university engagement
- Invest in social skills training for employees lacking these backgrounds
- Create internal activities fostering teamwork and collaboration, particularly during AI adoption phases

For policymakers

- Support university programs promoting holistic student development

- Encourage public-private partnerships facilitating extracurricular activities
- Integrate soft skills assessment into quality assurance frameworks

5.4 Limitations and Future Research

Several limitations warrant consideration:

Sample Limitations Despite adequate size (N = 450), geographic concentration in Morocco limits generalizability. Cross-cultural replications would enhance external validity.

Temporal Considerations Cross-sectional design prevents causal inference. Longitudinal studies tracking students from university through career progression would strengthen causal claims.

Self-report bias All measures relied on self-reports, potentially introducing common method bias. Future research should incorporate objective performance measures and supervisor ratings.

Activity measurement University activity participation was measured retrospectively, potentially subject to recall bias. Prospective designs would improve measurement accuracy.

Future research directions

- Longitudinal studies examining long-term effects of university activities on career trajectories
- Experimental interventions assessing social skills training effectiveness
- Qualitative research exploring mechanisms linking specific activities to organizational outcomes
- Cross-cultural comparisons examining cultural moderators
- Investigation of optimal activity combinations and intensity thresholds

6 Conclusion

This research demonstrates that university life dynamics constitute critical determinants of organizational climate in AI-integrated companies. Sports participation and volunteering emerge as powerful positive predictors, developing social and behavioral skills essential for navigating technological transformation. Conversely, entrepreneurial orientations and limited extracurricular engagement present challenges requiring specific interventions.

As AI reshapes organizational landscapes, human competencies become increasingly valuable. Universities must recognize their role extending beyond academic knowledge transmission to encompass holistic development preparing graduates for complex, technology-driven work environments. Companies, in turn, should value and leverage diverse university experiences when building adaptive, resilient organizational cultures.

The "human factor" remains irreplaceable in AI era—university experiences cultivating this factor deserve strategic attention from educational institutions, employers, and policymakers alike. Investment in holistic student development represents investment in organizational futures capable of thriving amid technological disruption.

References

Astin, A.W.: Student involvement: a developmental theory for higher education. J. Coll. Stud. Dev. **40**(5), 518–529 (1999)

Astin, A.W., Vogelgesang, L.J., Ikeda, E.K., Yee, J.A.: How Service Learning Affects Students. Higher Education Research Institute, UCLA (2000)

Bandura, A.: Social Learning Theory. Prentice Hall, Englewood Cliffs, NJ (1977)

Becker, G.S.: Human Capital: a Theoretical and Empirical Analysis. University of Chicago Press (1964)

Brock, J.K.U., Von Wangenheim, F.: Demystifying AI: what digital transformation leaders can teach you about realistic artificial intelligence. Calif. Manag. Rev. **61**(4), 110–134 (2019)

Brynjolfsson, E., McAfee, A.: The Second Machine Age: Work, Progress, and Prosperity in a Time of Brilliant Technologies. WW Norton & Company (2014)

Celio, C.I., Durlak, J., Dymnicki, A.: A meta-analysis of the impact of service-learning on students. J. Exp. Educ. **34**(2), 164–181 (2011)

Chui, M., Manyika, J., Miremadi, M., Henke, N., Chung, R., Nel, P., Malhotra, S.: Notes from the AI Frontier: Applications and Value of Deep Learning. McKinsey Global Institute Discussion Paper (2018)

Davenport, T.H., Ronanki, R.: Artificial intelligence for the real world. Harv. Bus. Rev. **96**(1), 108–116 (2018)

Dwivedi, Y.K., Hughes, L., Ismagilova, E., Aarts, G., Coombs, C., Crick, T., et al.: Artificial intelligence (AI): multidisciplinary perspectives on emerging challenges, opportunities, and agenda for research, practice and policy. Int. J. Inf. Manag. **57**, 101994 (2021)

Eccles, J.S., Barber, B.L., Stone, M., Hunt, J.: Extracurricular activities and adolescent development. J. Soc. Issues. **59**(4), 865–889 (2003) https://psycnet.apa.org/doi/10.1046/j.0022-4537.2003.00095.x

Fifani, S., Rabhi, D., Qachar, A.: The role of soft skills in managerial innovation and organizational performance: a qualitative study. Sci. Cult. **11**(3), 1006–1013 (2026)

Foubert, J.D., Grainger, L.U.: Effects of involvement in clubs and organizations on the psychosocial development of first-year and senior college students. NASPA J. **43**(1), 166–182 (2006)

Fraser-Thomas, J.L., Côté, J., Deakin, J.: Youth sport programs: an avenue to foster positive youth development. Phys. Educ. Sport. Pedagog. **10**(1), 19–40 (2005)

Hmieleski, K.M., Baron, R.A.: Entrepreneurs' optimism and new venture performance: a social cognitive perspective. Acad. Manag. J. **52**(3), 473–488 (2009) https://psycnet.apa.org/doi/10.5465/AMJ.2009.41330755

Holt, N.L., Neely, K.C., Slater, L.G., Camiré, M., Côté, J., Fraser-Thomas, J., et al.: A grounded theory of positive youth development through sport based on results from a qualitative meta-study. Int. Rev. Sport Exerc. Psychol. **10**(1), 1–49 (2017)

Jarrahi, M.H.: Artificial intelligence and the future of work: human-AI symbiosis in organizational decision making. Bus. Horiz. **61**(4), 577–586 (2018)
Jones, S.R., Hill, K.E.: Understanding patterns of commitment: student motivation for community service involvement. J. High. Educ. **74**(5), 516–539 (2003)
Kniffin, K.M., Wansink, B., Shimizu, M.: Sports at work: anticipated and persistent correlates of participation in high school athletics. J. Lead.Ship Organ. Stud. **22**(2), 217–230 (2015)
Kuh, G.D.: High-Impact Educational Practices: What they Are, Who Has Access to them, and why they Matter. Association of American Colleges and Universities (2008)
Kuh, G.D.: What student affairs professionals need to know about student engagement. J. Coll. Stud. Dev. **50**(6), 683–706 (2009)
Naranjo-Valencia, J.C., Jiménez-Jiménez, D., Sanz-Valle, R.: Studying the links between organizational culture, innovation, and performance in Spanish companies. Rev. Latinoam. Psicol. **48**(1), 30–41 (2016)
OECD: Skills for Social Progress: the Power of Social and Emotional Skills. OECD Publishing (2015)
Pascarella, E.T., Terenzini, P.T.: How College Affects Students: a Third Decade of Research, vol. 2. Jossey-Bass (2005)
Patterson, M.G., West, M.A., Shackleton, V.J., Dawson, J.F., Lawthom, R., Maitlis, S., et al.: Validating the organizational climate measure: links to managerial practices, productivity and innovation. J. Organ. Behav. **26**(4), 379–408 (2005)
Rauch, A., Hulsink, W.: Putting entrepreneurship education where the intention to act lies: an investigation into the impact of entrepreneurship education on entrepreneurial behavior. Acad. Manag. Learn. Edu. **14**(2), 187–204 (2015)
Rubin, M., Denson, N., Kilpatrick, S., Matthews, K.E., Stehlik, T., Zyngier, D.: "I am working-class": subjective self-definition as a missing measure of social class and socioeconomic status in higher education research. Educ. Res. **43**(4), 196–200 (2014)
Schneider, B., Ehrhart, M.G., Macey, W.H.: Organizational climate and culture. Annu. Rev. Psychol. **64**, 361–388 (2013)
Wilson, H.J., Daugherty, P.R.: Collaborative intelligence: humans and AI are joining forces. Harv. Bus. Rev. **96**(4), 114–123 (2018)

Managing Cultural Risk in GenAI Marketing: A Genre- and CDA-Informed Evidence Across USA-Targeted and Morocco-Targeted Market Framings

Khalil Hamidi, Hafssa Laabdi, and Mouad Mbeker

Abstract Generative AI is increasingly used to draft localized marketing copy, yet culturally non-neutral outputs can create trust, authenticity, and governance risks for organizations. In fact, there is a scarcity of genre-sensitive evidence on standardized web slots realized through controlled prompting across market framings despite GenAI's growth in terms of marketing research. The present study examines GenAI tourism website copy across two prompt-defined target framings, namely USA-targeted versus Morocco-targeted, anchored in genre theory and a CDA-informed discourse lens. The study opts for Hofstede and Hall's cultural dimensions heuristically as cultural sensitizing frameworks with McSweeney's critiques as a methodological guardrail. A total of 24 English texts was generated with the same prompt protocol and analyzed using a three-pass workflow that combines move/genre analysis of stance cues, culture cues, tourism cues, and CSR legitimacy cues. The findings revealed a mix of variable surface patterns and stable skeleton across both market framings. Move repertoires stay genre-faithful, while cross-framing variation appears in rhetorical realization, especially cue patterning and stance layering. On the one hand, USA-targeted outputs tend to systematically give the edge to explicitness cues and risk governance, while Morocco-targeted outputs do so for relational positioning in openings. On the other hand, CSR accountability staging varies, with ethical positioning sometimes appearing without comparably explicit monitoring/reporting cues. Taken together, the present study contributes a replicable auditing procedure and checklist-style checkpoints for responsible GenAI-assisted localization and calls for greater attention to conduct more future work across additional languages, models, and human reception measures.

K. Hamidi (✉) · H. Laabdi · M. Mbeker
Applied Language and Culture Studies Laboratory, Faculty of Letters and Humanities, Chouaib Doukkali University, El Jadida, Morocco, El Jadida, Morocco
e-mail: Hamidi.k816@ucd.ac.ma

H. Laabdi
e-mail: Laabdi.h367@ucd.ac.ma

M. Mbeker
e-mail: Mbeker.mouad@ucd.ac.ma

A. Bouziane and K. Bouziane (eds.), *AI, Management, and Culture: Shaping Organizations in the Digital Age*, Studies in Computational Intelligence 1291,
https://doi.org/10.1007/978-3-032-27882-1_7

Keywords Generative AI · Tourism website copy · Move/genre analysis · CDA · USA-targeted · Morocco-targeted

1 Introduction

Generative AI is quickly becoming part of organizational communication work, drafting website copy, campaign slogans, product narratives, customer-facing micro-texts, and even brand voice templates. Yet the role of generative AI in marketing communication has not received enough scholarly attention to its rapid uptake, creating a timing problem for both researchers and practitioners who must decide what good AI-assisted corporate communication should look like in culturally diverse markets (Kshetri et al. 2023). This timing problem is not merely technical. Corporate texts are cultural artifacts and they encode assumptions about audiences, authority, credibility, relationality, and emotion that can vary systematically across cultural contexts (Hall 1976; Hofstede 2001). When organizations use AI to scale or accelerate text production, the question becomes less "Can AI write?" and more "What cultural model of communication is being reproduced, normalized, or erased by AI-generated corporate discourse?" (Lu et al. 2025; Naidoo and Chadha 2025).

This question is especially salient in tourism communication, where language is not only informative but constitutive. Tourism texts help shape how places are imagined, desired, and evaluated. Besides, tourism discourse organizes gaze and expectation through communicative framing; as Dann memorably puts it, the "phrase precedes the gaze" (Dann 1996, p. 21). Tourism promotion therefore operates as a semiotic practice that guides attention, value, and emotional stance (Urry 2002), while also intersecting with long-standing debates about authenticity, staged experience, and representation (MacCannell 1976; Wang 1999). At the same time, tourism branding research shows how destination images are constructed through both cognitive (functional) and affective (psychological/holistic) components (Baloglu and McCleary 1999; Echtner and Ritchie 1993), and how branding strategies seek to stabilize distinctive place meanings in competitive markets (Blain et al. 2005; Pike 2002). If tourism discourse is already a managed cultural performance, then AI-generated tourism copy becomes a particularly revealing site for examining how culture enters corporate communication through choices about directness, politeness, collectivizing appeals, authority claims, emotional warmth, and implied relationships with audiences.

Importantly, corporate tourism communication is increasingly delivered through digital genres, where meaning is shaped not only by wording but also by the communicative slot and interface logic in which text appears. Genre, in applied linguistics, is treated as socially recognized communicative actions organized around shared purposes (Swales 1990) and highlights how professional communities produce patterned rhetorical moves to achieve institutional aims (Bhatia 1993). In web communication specifically, genres are often dual, that is, a website is not simply a text but also a medium with navigational and interactive affordances that influence how discourse is read and used (Askehave and Nielsen 2005). This matters because AI-generated corporate discourse is rarely free writing; it is typically generated into recognizable, high-stakes slots such as, homepage hero sections, About pages, CSR/sustainability statements. Each slot carries different expectations about voice, authority, and relationship-building which offers a controlled way to study cultural patterns in AI-generated corporate communication through a humanities lens (Bhatia 1993; Swales 1990).

Cross-cultural research already shows that culture influences how online corporate communication is received and trusted. Studies of e-service contexts indicate that cultural patterns can shape perceptions of website design and credibility, affecting trust and user response (Cyr 2008). Similarly, work on website localization argues that culturally mismatched web content can harm brand evaluation, and that culturally customized content matters, while standardized approaches can be problematic (Singh et al. 2005). These studies are highly relevant, but they largely examine human-produced web communication or localization as a managerial problem. Meanwhile, an emerging line of research on AI systems argues and provides initial evidence that AI-generated language may reflect systematic cultural tendencies (Lu et al. 2025), and research on culturally responsive chatbots emphasizes that cultural misalignment may undermine trust, reciprocity, and engagement in human–AI interaction (Naidoo and Chadha 2025). Another adjacent stream uses cultural dimensions, especially that of Hofstede, to guide AI-assisted persona generation and tests how cultural frameworks can scaffold persona design and evaluation (Avis et al. 2025). Taking together these strands suggest a pressing question, if AI-generated language carries cultural tendencies (Lu et al. 2025) and culture matters for online trust and brand communication (Cyr 2008; Singh et al. 2005), what happens when organizations use AI to generate corporate tourism discourse across culturally distinct contexts?

The gap this study addresses lies at the intersection of these literatures. We have (a) robust humanities-oriented work on tourism discourse and the semiotics of promotion (Dann 1996; Huang and Rust 2018; Urry 2002), (b) applied linguistics research on professional and digital genres (Askehave and Nielsen 2005; Bhatia 1993; Swales 1990), (c) cross-cultural studies of online trust and localization (Cyr 2008; Singh et al. 2005), and (d) emerging evidence that generative AI language may systematically pattern by cultural tendencies and that cultural responsiveness is a key challenge in AI-mediated communication (Lu et al. 2025; Naidoo and Chadha 2025). What remains comparatively underdeveloped is a focused, genre-sensitive, qualitative analysis of AI-generated corporate website discourse that (a) treats the texts as cultural-rhetorical artifacts not just content, (b) compares culturally distinct contexts through an explicit cultural-dimensions lens, and (c) does so in a domain where promotional language is known to shape perception and desire tourism (Dann 1996; Urry 2002). In other words, existing work helps us anticipate that culture matters and AI do patterns culturally, but we still lack a close reading of *how*, at the level of rhetorical moves, stance, implied relationships, and values, AI-generated corporate tourism genres may differ across cultural contexts.

To operationalize culture without reducing it to stereotypes, this study adopts cultural dimensions as heuristics rather than deterministic national essences. Hofstede's model (Hofstede 2001) remains influential in management and intercultural research, but it has also been heavily criticized for strong assumptions about national cultures and causal claims (McSweeney 2002). This critique is not a reason to abandon cultural dimensions, it rather motivates careful use, treating dimensions as interpretive resources, triangulating with discourse evidence, and avoiding claims that equate "country" with a single homogeneous culture (McSweeney 2002). To enrich interpretation at the level of communicative style, Hall's high−/low-context framework (Hall 1976) is also relevant for examining explicitness, relationality, and indirectness in corporate address across contexts.

Accordingly, this study investigates AI-generated tourism website copy designed for two prompt-defined target-audience contexts "USA-targeted" and "Morocco-targeted", which are treated as market framings (i.e., prompted audience framings) rather than empirical measures of national culture through a prompt-based, genre-controlled design. The analysis focuses on three common corporate website genres that are both rhetorically distinctive and feasible under tight timelines: homepage hero copy, About/brand story, and a CSR/sustainability statement. These genres are widely used in tourism branding to establish attractiveness and credibility (Blain et al. 2005; Pike 2002) and are well-suited for discourse-analytic comparison because they differ in communicative purpose and expected voice (Askehave and Nielsen 2005; Bhatia 1993). Using Critical Discourse Analysis as a guiding sensibility, attentive to how discourse constructs identities, relations, and values within institutional contexts (Wodak and Meyer 2009), the study examines how AI-generated texts may encode culturally patterned assumptions about authority, community, risk, authenticity, hospitality, and responsibility.

The study is guided by three research questions:

RQ1: How do AI-generated tourism website genres differ rhetorically across USA-targeted versus Morocco-targeted prompt conditions?
RQ2: What culture-linked discursive cues recur across target-audience contexts, and how are these cues shaped by genre?
RQ3: How do outputs construct authenticity and responsibility (CSR) claims across these two prompt-defined target-audience contexts?

This work contributes to works at the intersection of AI, management, and culture in three ways:

It offers a genre-sensitive discourse analysis of AI-generated tourism website copy across two target-audience prompt conditions (USA-targeted vs. Morocco-targeted); It develops a pluralized, non-essentialist cultural-dimensions lens as a heuristic for interpreting AI-mediated corporate discourse; It provides a transparent prompt-based design template (fictional brand, standardized prompts, archived outputs) that can be replicated or extended in future GenAI discourse studies.

By foregrounding genre and discourse, this study contributes a humanities-oriented approach to the theme of AI, management, and culture. It treats AI-generated corporate communication as a cultural-semiotic practice rather than a neutral productivity tool. In doing so, it also offers a practical implication. So, if AI-generated language tends to exhibit cultural patterns (Lu et al. 2025), then the principle of one-size-fits-all that generates brand copy can risk cultural flattening and trust loss in localized markets (Cyr 2008; Singh et al. 2005). The study therefore argues for culturally reflexive prompting and evaluation practices grounded in genre expectations and discourse evidence, informed but not governed by cultural dimensions frameworks (Hofstede 2001; McSweeney 2002).

2 Literature Review and Theoretical Framework

2.1 Culture, Management, and Communication: Using Dimensions without Essentializing

Cross-cultural management and organizational research have long treated culture as consequential for how coordination, leadership, and legitimacy are enacted and often operationalized through comparative dimensional frameworks (Hofstede 2001; House et al. 2004). Hofstede's model remains influential in management research and practice because it offers a parsimonious vocabulary for comparing value orientations (Hofstede 2001), and it has been widely applied to international branding and advertising to explain systematic differences in persuasive style and consumer meaning-making (De Mooij and Hofstede 2010). At the same time, Hofstede's model has been strongly criticized for over-generalization and for making weak causal claims about national cultures (McSweeney 2002). A complementary tradition is Hall's low/high context framework (Hall 1976), that links cultural patterning to degrees of implicitness/explicitness and reliance on shared background knowledge which is a relevant lens for corporate copy and tone.

To reduce dependence on a single framework, Trompenaars and Hampden-Turner's model (Trompenaars and Hampden-Turner 1997) is also useful, particularly because it targets relational and communicative dilemmas routinely activated in corporate discourse. Their dimensions map tensions such as universalism versus particularism, communitarianism versus individualism, neutral versus emotional communication, diffuse versus specific involvement, and achievement versus ascription status (Trompenaars and Hampden-Turner 1997). These concepts provide discourse-relevant cues for analyzing how organizations script relationships with audiences, for example, whether credibility is built through rule-like assurances, relational warmth, or status claims.

At the same time, dimensional approaches are not unproblematic. McSweeney's critique (McSweeney 2002) questions the plausibility of treating "the nation" as culturally uniform and warns that claims about national culture can become circular when national uniformity is presupposed rather than demonstrated. This critique is methodologically important for discourse work as it encourages treating cultural dimensions as heuristic lenses rather than deterministic explanations, and it discourages reading a country label as equivalent to a homogeneous cultural essence (McSweeney 2002). A further rationale for pluralizing the cultural lens comes from the GLOBE program, which conceptualizes culture through both practices, meaning the way things are done, and values, meaning the way things should be done, measured at multiple levels like, society and organization (House et al. 2004). This practices/values distinction is especially productive for corporate communication analysis because corporate texts, notably CSR genres, often perform precisely the "should be" layer of cultural morality and organizational identity.

2.2 *Cross-Cultural Corporate Websites: Localization, Design, and Trust*

Research on cross-cultural web communication demonstrates that websites are not culturally neutral containers. Cyr's model (Cyr, 2008) shows that the relationship between website design (information, navigation, visual design) and outcomes such as trust, satisfaction, and online loyalty had not previously been modelled across cultures and provides evidence that these relationships can vary by country context. This supports a central premise for the present study. When corporate communication is delivered online, cultural context can shape which cues are perceived as credible, usable, and trustworthy.

Localization research reinforces that cultural fit is more than translation. Singh, Zhao, and Hu (2005) provide evidence that culturally unique web styles emerge online and explicitly caution against launching standardized or machine-translated websites for global audiences. Gevorgyan and Manucharova (2009) similarly define cultural adaptation on the web as integrating core cultural norms into the structure/format of online messages directed at a group, while treating the presumed benefits of adaptation as an empirical issue rather than an automatic gain. Together, these studies justify analyzing tourism website copy as culturally consequential corporate discourse, while also positioning localization as a high-stakes managerial and communicative activity, which is relevant as generative AI is increasingly used to accelerate content production.

2.3 *Genre and Digital Genre: Why Website Slots Matter?*

To keep the inquiry humanities-oriented, the analysis is anchored in genre theory. Swales (1990, p. 58) defines genre as a class of communicative events whose members share communicative purposes, which he treats as "privileged" in defining their prototypical structures and styles, emphasizing the social organization of texts and the patterned realization of purpose. Bhatia (1993, p. 49) extends this perspective in professional contexts, defining genre as a recognizable communicative event characterized by shared purposes and conventionalized constraints, embedded in specific professional and institutional cultures, while also noting how expert writers can strategically exploit conventions to accomplish private intentions within socially recognized purposes. This framing supports treating corporate tourism pages as institutional discourse rather than merely content, and it legitimizes close analysis of rhetorical moves, stance, and credibility-building as genre performances.

Digital genres complicate genre because the medium shapes meaning. Askehave and Nielsen (2005) argue that web genres require a model that captures text and medium simultaneously, proposing a two-dimensional view that integrates the navigating mode with communicative purpose, move structure, and rhetorical strategies. Building on this framework, this study treats website elements such as the homepage hero, About/brand story, and CSR statements as recognizable digital genre slots with different communicative purposes, audience positioning, and action orientations.

2.4 *Tourism Discourse and Branding: Gaze, Authenticity, and Destination Image*

Tourism is a particularly revealing domain for analyzing AI-generated corporate discourse because tourism promotion is semiotic. It does not simply describe places but helps organize how they are imagined and valued. Urry (2002, p. 10) emphasizes that meaningful and profitable tourist gazes are socially produced; clear markers must be provided, and people must learn how, when, and where to gaze. In Dann's account (Dann 1996, p. 64), tourism communication is frequently unilateral and persuasion-oriented; tourism language can function as a monologue and is linked to the art of persuasion. Building on these accounts, tourism copy can be read as a form of cultural scripting that preconfigures desire, legitimacy, and expectation.

Authenticity debates sharpen the stakes of such scripting. MacCannell's staged authenticity (MacCannell 1976, pp. 92–93) explains how modern tourism relies on structured front- and back-region arrangements that transform authenticity into a staged social effect rather than a naturally given reality. Wang (1999) reframes authenticity by proposing existential authenticity as an alternative source of authenticity in tourism, regardless of whether toured objects are objectively authentic, and distinguishes intra-personal and inter-personal dimensions. These concepts are highly relevant for corporate tourism discourse, where promises of real culture, local experience, and ethical encounter are recurrent legitimacy strategies.

Destination image and branding research provides an explicit bridge to the management dimension of the study. Echtner and Ritchie (1993) argue that destination image includes both attribute-based components and holistic impressions, and that image can be conceptualized along functional–psychological and common–unique continua. Baloglu and McCleary (1999) define destination image as emerging from both cognitive evaluations (beliefs/knowledge about attributes) and affective evaluations (feelings/attachment), which combine into an overall image. Pike's review (Pike 2002) of destination image studies further indicates the breadth of the research area and the centrality of image as an organizing construct in tourism research. Branding-focused work defines destination branding as a set of marketing activities that create identifiers, convey expectations of memorable experiences, consolidate emotional connections, and reduce search costs and perceived risk, collectively shaping destination image and influencing destination choice (Blain et al. 2005). City branding research similarly argues that branding is an appropriate way to describe and implement city marketing, highlighting the conceptual shift from promotion-only approaches to broader identity construction (Kavaratzis 2004). These strands support analyzing tourism website genres as institutional discourse where identity, credibility, and emotional attachment are actively managed.

2.5 Generative AI in Marketing Communication: Managerial Urgency and Cultural Risk

Management and marketing research increasingly frames AI, including generative AI as a transformative development for marketing activities, while also noting a lag between organizational adoption and scholarly attention. Kshetri et al. (2023) explicitly argues that there is little research on how these tools are being used and on their effects on marketing activities, outcomes, and firm performance, positioning the topic as an emerging research agenda. In the advertising domain, Kietzmann et al. (2018) describe how marketers are turning to AI to transform large flows of consumer data into valuable consumer insight across the consumer journey, while also acknowledging risks and governance concerns. Huang and Rust (2018) propose that AI in service develops through four intelligences namely, mechanical, analytical, intuitive, and empathetic in a predictable sequence, and discuss how organizations decide between humans and machines in service provision, foregrounding the managerial implications of AI adoption for customer-facing communication and experience.

The cultural risk becomes sharper in light of evidence that generative AI is not culturally neutral. Lu, Song, and Zhang (2025) show that generative AI models trained on inherently cultural textual data exhibit cultural tendencies, and they caution against assuming language choice is neutral in deployment. This intersects with research on culturally responsive AI systems. Naidoo and Chadha (2025) argue that failures to recognize and respond to cultural differences can erode trust, reduce engagement, and undermine legitimacy, and they propose culturally responsive frameworks grounded in field evidence. Practical operationalization work also exists. Avis et al. (2025) evaluates AI-assisted persona generation guided by Hofstede's dimensions, illustrating both the appeal of cultural scaffolding and the need for careful evaluation to avoid oversimplification. Finally, cross-cultural evaluation of AI-generated advertising suggests that perception and engagement can vary by cultural context, reinforcing that cultural adaptation is also a reception problem (Diaz and Stöckl 2025).

2.6 Integrated Conceptual Framework for the Present Study

Synthesizing these strands, the present study conceptualizes AI-generated tourism website copy as culturally loaded, genre-constrained corporate discourse shaped by managerial pressures to scale communication. The framework has four interlocking components. To begin with, ***digital genre as analytical anchor***, where website sections are treated as professional/digital genres with distinct communicative purposes and patterned rhetorical moves (Bhatia 1993; Swales 1990), and are analyzed with attention to web-mediated text-and-medium dynamics (Askehave and Nielsen 2005). Moreover, ***tourism meaning making***, where outputs are interpreted through tourism discourse theory (persuasion, gaze, markers) and authenticity debates, which explain how tourism texts construct the real, the local, and the desirable (Dann 1996; MacCannell 1976; Urry 2002; Wang 1999). Additionally, ***branding and destination image as managerial discourse***, where the analysis draws on destination image and branding research to examine how AI-generated copy performs differentiation, builds emotional connection, and supports risk reduction through discourse (Baloglu and McCleary 1999; Blain et al. 2005; Echtner and Ritchie 1993; Kavaratzis 2004; Pike 2002). Finally, ***Culture as heuristic (pluralized and reflexive)***, where cultural dimensions are used to sensitize analysis to patterned differences in stance, relationality, authority, and explicitness, while explicitly avoiding national-cultural essentialism by drawing on critiques and alternatives (De Mooij and Hofstede 2010; Hall 1976; Hofstede 2001; House et al. 2004; MacCannell 1976; Trompenaars and Hampden-Turner 1997).

Within this framework, the core gap becomes more precisely defined: cross-cultural website research explains why trust/localization matters (Cyr 2008; Gevorgyan and Manucharova 2009; Singh et al. 2005), tourism and branding research explains how destinations are discursively produced (Blain et al. 2005; Dann 1996; Urry 2002), and emergent AI research suggests that GenAI outputs can carry cultural tendencies and require cultural responsiveness (Lu et al. 2025; Naidoo and Chadha 2025). There is no work here offering a genre-sensitive, discourse-analytic account of AI-generated tourism website genres across cultures; existing AI culture work focuses on chatbots, personas, or general web communication rather than destination web copy. Despite rapid adoption of GenAI for localized marketing copy, there remains limited genre-sensitive, cross-cultural work that audits how standardized move skeletons are rhetorically realized under controlled prompting, especially in tourism website micro-genres.

3 Methodology

3.1 Research Design and Orientation

This study adopts a qualitative, humanities-oriented design to examine how generative AI performs culture in corporate tourism website discourse. The analysis combines genre analysis and Critical Discourse Analysis CDA. Genre analysis is used to examine how corporate texts realize communicative purposes through patterned rhetorical moves within recognizable website genres (Bhatia 1993; Swales 1990). CDA functions less as a fixed method and more as a problem-oriented interpretive stance for examining how these texts construct identities, relationships, and values within institutional communication (Wodak and Meyer 2009). Following Wodak and Meyer, CDA is treated here as an *eclectic* and methodologically plural family of approaches rather than a single canon, so it serves as an interpretive orientation guiding what is attended to in the texts, not a fixed procedure. Together, these approaches support close, comparative reading of AI-generated corporate discourse across target-audience contexts.

3.2 Data and Corpus Construction: Prompt-Based Comparative Design

The dataset consists of a prompt-generated corpus of tourism website texts produced under controlled conditions. The aim is not to model population-level national discourse, but to examine how ChatGPT (OpenAI, GPT-5.2; accessed 31 December 2025) performs culturally inflected corporate discourse when instructed to write for different target-audience contexts. Texts were generated via the ChatGPT interface using only the stated prompts (no additional system instructions or external tools). For each condition, each output was generated in a fresh chat thread to minimize cross-prompt contamination and conversational memory effects and copied verbatim without post-editing; prior outputs were not shown to the model.

Contexts Two target-audience contexts are compared (USA-targeted vs. Morocco-targeted), operationalized as prompted market framings rather than as direct proxies for national cultures; visitor type and travel scenario were held constant across contexts.

Genres Three website genres are sampled because they are widespread in corporate tourism communication and rhetorically distinct:

Homepage hero copy (first-contact promise + affective framing + action orientation)
About/brand story (identity + legitimacy + relationship-building)

CSR/sustainability statement (moral positioning + responsibility + stakeholder stance)

This genre selection is theoretically grounded in the view that web-mediated genres are simultaneously text and medium, where navigation, linking, and reading paths shape meaning-making (Askehave and Nielsen 2005).

Fictional Brand Scenario (Standardization and Ethics) The corpus is generated using a single fictional tourism operator *Atlas Echo Trails*. A fictional brand is used to standardize the scenario across conditions and reduce confounds associated with real brands namely, existing reputations, recognizable taglines, or inadvertent reproduction of existing website language. This choice also avoids ethical and legal issues tied to transforming proprietary corporate copy. The fictional scenario therefore functions as a methodological device that supports comparability, transparency, and replicability in a prompt-based design.

Sampling Logic and Corpus Size A repeated-generation strategy is used to capture controlled variation without producing an unmanageable corpus. For each condition (2 contexts × 3 genres), four outputs are generated, yielding a corpus of 24 texts (2 × 3 × 4). Each output was generated in a fresh chat thread using the same prompt template for that condition. All texts were generated on 31 December 2025 using platform-default generation settings, with no manual temperature or top-p adjustments beyond the prompt text. Because platform models can be updated and outputs are stochastic, rerunning the same prompts at a later date may not reproduce identical texts.

Accordingly, the study prioritizes procedural reproducibility (the same protocol, controls, and analytic workflow) rather than strict output replication. To support transparency, full verbatim prompts (as entered) and outputs are archived in the Supplementary audit-trail workbook, while Appendix A provides a condensed overview of the prompt protocol and the move-presence legend used in analysis. The resulting corpus (N = 24) is intentionally compact to enable close discourse/genre analysis and cross-case comparison; analytic saturation was judged based on recurring genre and stance patterns rather than frequency counts. Findings are specific to ChatGPT (GPT-5.2) as accessed on 31 December 2025 and should not be generalized to all generative AI systems.

3.3 *Prompt Protocol and Generation Procedure*

To maximize comparability across contexts and genres, prompts follow a shared template with only two controlled variables namely, target-audience context (USA-targeted vs. Morocco-targeted) and genre slot (hero/About/CSR).[1] Target-audience

[1] Hero/About/CSR are used as a shorthand for homepage hero copy, About/brand story, and CSR/sustainability statement.

context was operationalized by specifying the same visitor type and travel situation in both conditions (prospective eco-tour clients planning a 2–5-day High Atlas trip), while varying only the target-market framing (USA-targeted vs. Morocco-targeted) in the prompt (see Appendix A for the prompt protocol summary; full verbatim prompts are in the Supplementary audit-trail workbook). The brand scenario, service features, and tone constraints remain constant across conditions.

The protocol includes, a ***Fixed scenario***, in other words, a Marrakech-based eco-tour operator offering 2–5-day High Atlas trips, small groups (max 10), local guides, clear safety/logistics planning, and stated sustainability practices (local partnerships, low-waste habits, leave-no-trace briefing). ***Fixed length ranges***, across Homepage hero (45–70 words), About/brand story (120–180 words), and CSR statement (140–220 words). ***Fixed tone constraints***, where a consistent baseline tone was specified as professional, warm, and credible. ***Anti-stereotyping instruction (ethical and analytic control)***, in which the explicit clause "Avoid stereotypes and exoticizing language; represent local communities respectfully," is included in the prompts reflecting concerns that cultural responsiveness is not automatic and must be treated as a design/evaluation issue. This constraint is methodologically supported because what counts as polite, persuasive, or trustworthy can shift across markets, and cultural misalignment can reduce legitimacy and trust in AI-mediated communication (Naidoo and Chadha 2025). Eventually, ***Documentation***, a condensed overview of the prompt protocol and move-presence legend is provided in Appendix A. The full audit-trail workbook (verbatim prompts/outputs, segment-level coding table, codebook, themebook, and summaries) is provided as Supplementary material.

3.4 Language Choice and Control of Confounds

All outputs are generated in English for both contexts as a methodological control to minimize the risk that observed cross-cultural differences stem from translation effects or language-specific generation behavior. This is particularly important given evidence that generative AI can display systematic cultural tendencies and that language conditions can shape outputs (Lu et al. 2025). This choice also defines a clear limitation where Moroccan tourism communication often circulates in Arabic and French as well as English. This responds to Lu, Song, & Zhang's warning as discussed in (Sect. 2.5). The study therefore analyses culture-in-discourse under a controlled English-output condition and explicitly brackets multilingual practice as a limitation and future extension.

3.5 Analytical Procedure

Analysis is conducted in three recursive passes, moving between the corpus, coded extracts, and interpretive claims. This recursive organization reflects Braun and

Clarke's emphasis that thematic analysis is not linear and involves moving back and forth across phases as coding and themes are refined (Braun and Clarke 2006). The primary unit of analysis is each generated text (N = 24); a secondary unit of analysis is salient discursive segments, stretches of text realizing a single communicative move, and genre moves within each text.

Given that coding and memoing were conducted by a single researcher using a structured codebook and a manual spreadsheet-based workflow; reflexive memos were used to surface and document interpretive decisions and potential biases, consistent with qualitative discourse-analytic practice (Braun and Clarke 2006). Thematic analysis is used here primarily as an organizing device to structure patterned meanings across the corpus, while close discourse and genre analysis remain the primary interpretive engines. Findings are interpreted as patterns in model-generated discourse under specified prompting conditions, not as direct evidence of national communication behavior. The analysis therefore reports patterns in model performance under specified prompting constraints.

Pass 1: Genre Move Analysis Each text is analyzed as a genre performance, focusing on how communicative purpose is realized through recurrent moves and rhetorical sequencing (Bhatia 1993; Swales 1990). Following Swales, communicative purpose is treated as the privileged anchor for identifying and comparing move structure within each genre. For each genre, texts were segmented into a consistent set of move-based units to support within-genre comparison; where a move was implicit, it was coded as an implicit realization rather than omitted. Implicit/absent realization of expected moves was recorded in a move-presence audit matrix; Appendix A reports the move-presence legend and an illustrative excerpt, while the full matrix is available in the Supplementary audit-trail workbook. Segment rows were coded by the dominant observed move, with secondary/implicit expected moves recorded separately; slot misalignment was treated as sequencing variation. An expected move was coded as implicit only when its function was materially realized (at least clause-level) within another segment and contributed meaningfully to the communicative purpose; otherwise, it was treated as absent and noted in memoing. Differences are then compared across USA-targeted versus Morocco-targeted outputs within each genre, to be more specific, hero versus hero, About versus About, and CSR versus CSR to maintain functional comparability.

Pass 2: Thematic Coding across the Corpus A reflexive thematic analysis procedure is used to organize patterned meanings across the dataset, following Braun and Clarke's guidance (Braun and Clarke 2006) that analysis is iterative and recursive rather than linear. Coding begins with broad descriptive codes (stance, audience positioning, authenticity claims, responsibility framing) and then consolidates into themes. Coding followed hybrid logic. Deductive codes from the coding grid were applied first, and inductive sub-codes were developed during iterative memoing (Braun and Clarke 2006). Any counts derived from theme-presence summaries are used descriptively to map coverage and patterning across the corpus (as a transparency check), not as inferential evidence or a substitute for interpretive qualitative analysis.

Pass 3: Culture-Lens Interpretation Interpretation draws primarily on Hofstede (2001) and Hall (1976) as sensitizing lenses for communicative style; Trompenaars and GLOBE are used selectively to contextualize patterns in relational and CSR discourse (House et al. 2004; Trompenaars and Hampden-Turner 1997), reducing reliance on any single model. Throughout interpretation, McSweeney's critique (McSweeney 2002) functions as a methodological guardrail. Cultural frameworks inform reading but are not used to essentialize national contexts or convert discourse patterns into deterministic national traits.

3.6 Coding Grid: Operationalizing Genre and Culture in Discourse

Coding integrates genre structure and culture-sensitive discourse cues, using a structured coding grid; the full codebook and operational definitions are provided in the Supplementary audit-trail workbook. The analysis focuses on five domains which are Genre moves and rhetorical sequencing that deals with promise, credibility work, call-to-action (Bhatia 1993; Swales 1990), Interpersonal stance and audience positioning that focuses on pronouns, modality, directness, politeness, emotional warmth/formality, Culture-linked communicative cues operationalized heuristically, for instance, individual versus collective framing, authority positioning, uncertainty/risk management, and explicitness/implicitness (Hall 1976; Hofstede 2001).

For example, uncertainty/risk management was coded through explicit safety reassurance moves and modality choices of certainty and hedging; Culture-linked cues are treated as heuristic, text-anchored indicators and were assigned only when a cue was foregrounded as the segment's primary rhetorical signal. Segments with no salient primary cues were coded as CC_NONE; full indicators and decision rules are documented in the Supplementary codebook/audit-trail workbook, Tourism meaning-making cues (MacCannell 1976; Wang 1999), and CSR legitimacy cues (House et al. 2004; Wodak and Meyer 2009). This operationalization enables systematic comparison across genres and target-audience contexts while keeping cultural interpretation anchored in textual evidence (McSweeney 2002). Also, comparisons are made at the level of patterned discursive tendencies across prompt conditions, not at the level of individual texts.

3.7 Trustworthiness, Transparency, and Reflexivity

Credibility is strengthened through 6 major components, First, a *full audit-trail workbook* (segment-level coding table, move-presence audit matrix, theme-presence summaries, and stability check) is available as supplementary material for transparency. Second, *recursive coding and memoing*, consistent with Braun and Clarke's emphasis (Braun and Clarke 2006) on iterative movement between data, codes, and themes. Third, an *intra-coder consistency check* where six texts (25%), one from each genre × context cell, were re-coded after a 5-day interval to assess coding stability at the segment level and the stability of theme boundaries. No major shifts in main themes occurred. Fourth, a *negative case analysis* where outputs that contradict emerging patterns are preserved and discussed. Fifth, a *framework pluralization* using multiple cultural models to reduce overfitting interpretation to a single dimension set (Hofstede 2001; House et al. 2004; Trompenaars and Hampden-Turner 1997). And eventually, *anti-essentialist caution interpretations* to avoid equating country label with homogeneous culture, in line with McSweeney's critique (McSweeney 2002).

3.8 Ethical Considerations

No human participants are involved; the dataset consists of AI-generated texts. Using a fictional brand scenario also avoids reproducing or transforming proprietary corporate texts and reduces the risk of inadvertent copying. Nonetheless, cultural representation is treated as ethically consequential. Analysis explicitly attends to stereotyping, cultural flattening, and overconfident cultural claims in generated corporate discourse, consistent with concerns that cultural responsiveness in AI-mediated systems requires deliberate design and evaluation rather than assumption (Naidoo and Chadha 2025).

4 Results

This section reports patterned differences in model-generated tourism homepage discourse produced under controlled prompting constraints. Findings are interpreted as patterns in model outputs under specified prompting conditions, rather than as direct evidence of national communication behavior. Where presence counts or implicit/absent labels are referenced, they are used descriptively to document coverage and comparability across texts.

4.1 Corpus Overview and Segmentation

The dataset comprises 24 generated texts (N = 24). Two target contexts (USA-targeted vs. Morocco-targeted) across three homepage genres (HERO, ABOUT/brand story, CSR/sustainability statement). Analysis follows the three recursive passes specified in the methodology, (a) move analysis, (b) thematic consolidation, and (c) culture-lens interpretation. For within-genre comparability, each text was segmented into move-based units. Segment rows were coded by the dominant observed move, while secondary/implicit expected moves were recorded separately. Where an expected move was materially realized inside another segment, it was coded as an implicit realization; if not realized, it was coded as absent and documented through memoing. These implicit/absent realizations were tracked in a move-presence audit matrix (Appendix A) to preserve an audit trail of genre expectations versus textual realization.

4.2 Genre Move Structure under Controlled Prompting

The first research question asks how genre-specific communicative purpose is realized through recurrent moves and rhetorical sequencing across the USA-targeted and Morocco-targeted outputs. Following the genre-move procedure described in the Method section, results are reported within genre (HERO vs. HERO, ABOUT vs. ABOUT, CSR vs. CSR) to maintain functional comparability. Across the corpus, results show high structural stability. For each genre, texts largely draw from a consistent move repertoire, with variation occurring mainly in (a) whether a move is realized explicitly or implicitly, and (b) how the move is rhetorically performed, rather than whether the communicative function is present at all.

HERO (Homepage HERO Copy) HERO texts consistently realize a four-move sequence:

H1: opening hook + positioning (what this is/who it's for)
H2: core offer + practical reassurance (trip basics, logistics, competence cues)
H3: responsibility practices (ethics/sustainability/local partnership, when present)
H4: closing payoff (final promise and/or invitation/next step)

Move presence is stable across the USA-targeted set (all four moves realized explicitly). In the Morocco-targeted set, two texts include implicit realizations rather than standalone segments: MOR-HERO-2 (H4 implicit) and MOR-HERO-3 (H3 implicit). Importantly, the communicative functions remain identifiable but are embedded within adjacent segments rather than signposted as distinct move units (see Appendix A).

ABOUT (Brand Story/Operator Identity) ABOUT texts display a stable move repertoire that builds identity and legitimacy through:

A1: brand identity/origin/values
A2: offer overview + audience positioning
A3: operational credibility (planning, safety, transparency, competence)
A4: responsibility/partnership practices (community/environment positioning)
A5: closing promise/invitation (final takeaway; conversion-friendly close)

Across the USA-targeted set, ABOUT texts realize the expected moves explicitly. In the Morocco-targeted set, MOR-ABOUT-1 codes A5 as absent, producing an ending that closes on identity/values and commitments without a distinct invitation/closing payoff (documented in Appendix A). This difference is treated as a move realization difference (not a genre mismatch), because the ABOUT communicative purpose remains intact.

CSR (Sustainability/Responsibility Statement) CSR texts typically stage:

C1: CSR stance/principle (what we believe)
C2: stakeholders/partnerships (who we work with; shared responsibility)
C3: environmental practices (impact reduction actions)
C4: transparency/accountability (monitoring, reporting, specificity)
C5: closing goal/long-term impact (future-facing commitment)

Move structure is highly stable across contexts, with one notable exception: MOR-CSR-2 codes C4 as absent, meaning the text emphasizes values and practices without a distinct accountability/reporting segment (captured in Appendix A). As with ABOUT, this is interpreted as variation in realization rather than absence of responsibility discourse altogether. Under controlled prompting, the model produces genre-faithful move structures in both target contexts. Differences appear primarily as explicit versus implicit realizations of expected moves and occasional absences of particular closing/accountability moves, rather than systematic breakdowns of genre structure.

4.3 Stance and Culture-Linked Communicative Cues (Descriptive Patterns)

The second research question examines how the outputs differ in interpersonal stance and culture-linked cues (as operationalized in the coding grid) across USA-targeted versus Morocco-targeted texts. Results below report coded distributions and typical realizations, with brief segment exemplars for transparency.

Credibility as the shared baseline, with genre-sensitive tone layering. Across the full corpus, credibility-building is the dominant primary stance (ST_CRED = 60/112 segments). This holds across contexts and genres, indicating that the outputs systematically prioritize competence and trustworthiness as a default.

Where contexts diverge is not in *whether* credibility is present, but in how it is layered with other stances:

In HERO, USA-targeted segments show more professional framing as the primary stance (ST_PRO: 4/16 USA-HERO vs. 1/16 MOR-HERO), while Morocco-targeted HERO shows more warmth/relational tone as primary (ST_WARM: 4/16 MOR-HERO vs. 0/16 USA-HERO).

In ABOUT, both contexts rely heavily on ST_CRED (12/20 each), with Morocco-targeted ABOUT showing slightly more warmth and "we"-voice as primary in some segments.

Illustrative cue-level contrast (HERO opening):

MOR-HERO-1_S1: "Walk the High Atlas with people who know it well." (warm/relational positioning)
USA-HERO openings tend to start with a more informational, operator-positioning frame (brand + offer + competence).

Risk governance and explicitness cues cluster in HERO and ABOUT. Two cue families recur strongly in the corpus:

Risk governance/uncertainty management (CC_RISK).
Explicitness/clarity signaling (CC_EXPL).

In HERO, CC_RISK appears as the primary cue at the same rate in both contexts (6/16 USA-HERO; 6/16 MOR-HERO), typically realized via safety planning, predictable logistics, and controlled group parameters. This pattern is consistent with the genre function of reducing perceived travel uncertainty quickly in the opening screen.

In ABOUT, CC_EXPL is frequent in both contexts (6/20 USA-ABOUT; 7/20 MOR-ABOUT) and often appears alongside credibility moves (operational transparency, what is handled, what the traveler can expect).

Exemplars (risk/explicitness realizations):

USA-HERO-1_S2 foregrounds "clear safety planning and logistics handled."
MOR-ABOUT-1_S3 similarly uses small-group + guide expertise + operational clarity to support credibility.
USA-ABOUT-1_S3: "From permits and transport to meals and lodging, plans are transparent and safety-forward."
MOR-ABOUT-1_S4: "From route selection to meals and overnight stops, every detail is planned with safety, comfort, and clarity…"

Relational cues are most visible in Morocco-targeted HERO, while community-benefit cues appear in both contexts. Relational framing (CC_REL) shows the clearest context separation in HERO:

CC_REL as primary cue is present in MOR-HERO (3/16) and absent in USA-HERO (0/16).

This does not mean USA-targeted texts lack relational language; rather, in HERO specifically, USA-targeted outputs more often prioritize risk/explicitness/competence cues as the primary coded signal in the opening segments, while Morocco-targeted HERO more often allows the hook to foreground relationship, as in the example "people who know it well", in addition to respect and exchange before logistics.

Exemplars (relational positioning outside HERO openings):

MOR-CSR-1_S4: "We see ourselves as guests as well as hosts. Fair collaboration…"
USA-ABOUT-1_S5: "Whether you're new to trekking or returning to the mountains…"

By contrast, community/collective benefit cues (CC_COLL) are common across CSR and frequently appear as primary cues in both contexts (7/20 USA-CSR; 8/20 MOR-CSR), typically realized through local benefit, partnership, and community support framing.

Reader positioning: "we" voice and direct address concentrate in ABOUT and closure moves. Two secondary stance patterns help explain how texts manage involvement:

ST_WE appears frequently as a secondary stance (38 segments overall), especially in ABOUT and CSR where organizational ethos is constructed through "we believe/ we do/we commit."

ST_AUD_YOU (direct address) is most concentrated in HERO and ABOUT (especially in payoff/closure segments), functioning as an engagement mechanism and soft call-to-action layer.

4.4 Authenticity Work and CSR Legitimation

The third research question focuses on how outputs construct (a) authenticity/meaning in tourism discourse and (b) CSR legitimacy, using the tourism-meaning codes, CSR legitimacy codes, and inductive theme buckets.

Authenticity and Meaning are built mainly through local grounding and responsible travel framing. Across both contexts, tourism-meaning coding shows that local grounding (TM_LOCAL) and adventure framing (TM_ADVENT) dominate HERO and ABOUT, while CSR pages foreground responsible travel meanings more strongly. A small but notable pattern appears in the inductive theme layer. The theme bucket TB7 (Authenticity & cultural exchange) occurs only in Morocco-targeted segments (3 instances) and not in the USA-targeted set. In practical terms, the Morocco-targeted outputs more explicitly lexicalize cultural exchange/relational authenticity as a discrete theme, while the USA-targeted outputs more often express authenticity-adjacent meaning through competence, responsibility, or local-guide credibility framing rather than an explicit exchange emphasis.

Exemplar (authenticity/relational positioning):

MOR-HERO-1_S1 "people who know it well" and MOR-ABOUT identity segments emphasize lived knowledge, respect, and exchange as part of the brand promise.

CSR legitimacy is primarily practice-based, with variable explicit accountability signaling. CSR legitimacy coding indicates that both contexts strongly emphasize:

Impact/ethics framing (CSR_IMPACT).
Leave-no-trace and low-waste practices.
Partnership/community benefit (CSR_PARTNER).

Transparency/accountability signaling (CSR_TRANSP) appears only in CSR pages (as expected) and is slightly more frequent in the Morocco-targeted CSR set (6 instances) than in the USA-targeted CSR set (4 instances). However, this does not eliminate variability at the move level. One Morocco-targeted CSR text was coded with the accountability/reporting move absent (C4 absent), meaning accountability was not staged as a distinct rhetorical unit even when responsibility practices were present. Boundary case (accountability staging) In MOR-CSR-2 express responsibility through partnerships and practices without a standalone accountability/reporting segment (C4 absent). For instance:

- MOR-CSR-2_S3: "We work through long-term partnerships with community hosts…"
- MOR-CSR-2_S4: "On the trail, we follow low-waste habits and provide a practical leave-no-trace briefing …"

By contrast, USA-CSR-2 stages accountability explicitly as a distinct unit:

- USA-CSR-2_S4: "Transparency matters to us. We explain logistics, risks, and expectations upfront …"

Exemplar (community benefit/ethics positioning):

CSR segments in both contexts repeatedly tie small-group operations to community benefit and long-term relationships, framing legitimacy through how we operate rather than only slogans.

Authenticity and CSR legitimacy are constructed through highly recurrent resources (local grounding, responsible travel, community benefit, practice-based ethics). Differences emerge mainly in the explicitness of cultural exchange theming (more overtly separated in MOR-coded themes) and in how consistently CSR outputs stage accountability as its own move.

4.5 Cross-RQ Synthesis

Taken together, the results suggest a consistent stable skeleton, variable surface pattern across the corpus. Under controlled prompting, the outputs largely preserve genre-faithful move repertoires within each homepage genre (HERO/ABOUT/CSR), while differences across USA-targeted versus Morocco-targeted texts emerge mainly through how moves are realized, whether explicit or implicit staging, relational or risk-governance emphasis, and tonal layering, rather than through systematic breakdown of genre structure. At the thematic level, both contexts draw on overlapping semantic resources like, credibility, operational transparency, responsibility practices, but distribute them differently across openings and closures, producing small but interpretable differences in interpersonal stance and in how authenticity/CSR legitimacy is performed.

5 Discussion

5.1 Genre Constraints Dominate Move Structure; Variation Appears as Rhetorical Strategy

A central interpretation is that genre constraints strongly condition the model outputs under controlled prompting. In Swalesian terms, genres are classes of communicative events organized around shared communicative purposes, which are "privileged" because they shape prototypical structures and stylistic choices (Swales 1990, p. 58). This helps frame the result. So, when the prompt stabilizes the communicative purpose, whether homepage hero, About, or CSR, the outputs tend to preserve a recognizable genre skeleton, even when surface realizations differ. This interpretation also aligns with web-genre works that operationalizes genre as a chain from communicative purpose → move structure → rhetorical strategies (Askehave and Nielsen 2005). In this framing, communicative purpose is not merely a label; it is the organizing principle that links what a text is for, to how it is staged, through moves, and how it sounds, through strategic rhetorical choices.

Importantly, genre instances do not require a rigid, invariant set of obligatory moves; instead, they can select structural elements from a shared move repertoire, leaving room for rhetorical variation (Askehave and Nielsen 2005). The audit trail captures this pattern: most texts realize the expected functional stages, while variation appears as (a) implicit realizations (functions embedded within adjacent segments) or (b) occasional absences of a distinct move that would be expected in the genre slot. Rather than forcing these into the segment grid, they were documented transparently in the move-presence audit (see Appendix A). This is also why the decision rule of coding each segment row by the dominant observed move while recording secondary or implicit expected moves separately remains methodologically solid. Move identity is ultimately functional, not merely formal, and genre analysis treats rhetorical options as flexible realizations within conventionalized structures (Bhatia 1993; Swales 1990).

Finally, because these are web-mediated texts, the pattern is especially plausible. Digital genres act not only as text but also as medium, where navigation and reading constraints shape production and reception (Askehave and Nielsen 2005). This helps explain why outputs converge on similar staging solutions across contexts even when interpersonal tone and emphasis vary. While move structure is largely stabilized by genre constraints, a CDA-oriented reading helps specify where variation concentrates, especially in how legitimacy and trust are constructed through stance, agency, and participant roles across the prompt-defined contexts.

5.2 CDA Lens: Agency, Legitimacy, and Representation

Using CDA as a sensitizing interpretive stance (Wodak and Meyer 2009), variation across market framings appears mainly in how the AI texts construct legitimacy and allocate agency within genre-faithful slots. Genre analysis points to a shared skeleton, while CDA helps interpret how different trust rationales and relationships are staged through stance and participant roles.

In the HERO and ABOUT genre slots, Morocco-targeted outputs more often ground legitimacy in relational authority and lived attachment, as illustrated in "people who know it well/by heart", and guides who know the mountains, as shown in "through daily life, not just maps", alongside relational-value cues such as "genuine exchange" and "respect everyday life". USA-targeted outputs more often ground legitimacy in managed competence and governance-coded reassurance, as in "clear safety planning and logistics handled", "seamless logistics", and "plans are transparent and safety-forward", positioning the traveler as benefiting from organizational control, in other words, focus on the trail, not the details.

In CSR, both framings reiterate small-group impacts, partnerships, low-waste practice, and leave-no-trace briefings, but USA-targeted CSR more strongly foregrounds transparency as governance, as demonstrated in "Transparency matters..." "explain logistics, risks, and expectations upfront", whereas Morocco-targeted CSR more often frames responsibility as fitting "the rhythm of mountain life" and exercising "restraint". These contrasts are interpreted as patterned tendencies in model outputs under prompt-defined framings not as national communication behavior, making it consistent with the "stable skeleton, variable surface" result where moves remain stable but legitimacy cues and agency positioning vary across contexts.

5.3 Culture-Lens Interpretation without Essentialism: Prompts as Target Framing, Culture as a Sensitizing Resource

The cross-context contrasts observed in this corpus are best interpreted as target-framing effects in model outputs, not as direct evidence of how Americans or Moroccans communicate. And for that reason, cultural frameworks are used as sensitizing lenses rather than as deterministic explanations. This approach is consistent with the logic of cultural-dimension work as heuristic patterning, while also acknowledging longstanding critiques that national culture models can overgeneralize and underplay within-country variation and contextual complexity (Hall 1976; Hofstede 2001; McSweeney 2002).

Relational warmth as a front-loaded strategy in Morocco-targeted HERO. One of the clearest context-linked differences appears in HERO openings. For instance, Morocco-targeted texts more often foreground a warm/relational stance as the primary frame, while USA-targeted HERO openings more often foreground professional/competence cues. Interpreted cautiously, this pattern can be described in terms of a stronger relational orientation in the Morocco-targeted framing, which resonates with cultural models that contrast more relationship-centered versus more task-centered communication preferences (Hall 1976; Hofstede 2001). Importantly, this is not treated as a claim about audiences; rather, it is evidence that the model, when instructed to target Morocco, selects relational language earlier in the text's rhetorical sequence.

This finding also aligns with web-localization research showing that culturally adapted websites often vary in how directly they address the user, how they establish trust, and how they balance relational versus transactional cues (Cyr 2008; Singh et al. 2003; Singh et al. 2005). In the data, the difference is not a different genre altogether (the move skeleton remains stable), but a difference in where relational framing is placed, often at the opening hook, before operational reassurance.

Explicitness and risk-governance cues are shared but deployed differently by move position. A second robust pattern is the prominence of risk governance and clarity/explicitness cues across both contexts, particularly in HERO and ABOUT. This convergence is not surprising in a tourism context; homepage discourse typically needs to reduce uncertainty and establish competence quickly. From a cultural-dimension perspective, uncertainty management can be treated as a general communicative resource that may be amplified under certain cultural framings (Hofstede 2001), but the findings suggest the more important point is distribution by move position. USA-targeted texts tend to foreground operational clarity and competence earlier and more explicitly; Morocco-targeted texts also include these cues, but more often layer them after or alongside relational framing.

These kinds of same resources and different sequencing results fit well with digital localization accounts. Localization is often less about wholly different content and more about how content is prioritized, grouped, and staged to match a target framing (Cyr 2008; Singh and Pereira 2005).

Culture-lens interpretation as pattern description, not cultural attribution. Because cultural frameworks can tempt analysts into explaining patterns too quickly, the study adopts a guardrail stance, the interpretation stays at the level of discourse patterns (warmth, direct address, explicitness, uncertainty management, community benefit cues) and avoids reifying them as stable national traits. This is particularly important given McSweeney's critique (McSweeney 2002) that national-culture scores can be treated as facts in ways that exceed what the data warrant. Where broader cultural models are invoked, they are used only to name plausible interpretive dimensions, not to assert audience psychology.

To further reduce essentialism, the discussion treats cultural explanations as plural and competing rather than singular. For example, where Hofstede's dimensions (Hofstede 2001) can frame a pattern in terms of individualism/collectivism or uncertainty avoidance, alternative culture frameworks such as, practice-versus-values distinctions, can also be relevant and remind us that culture is not one variable with one mechanism (House et al. 2004; Trompenaars and Hampden-Turner 1997).

Why this matters for GenAI: cultural tendencies as model behavior under prompts. Finally, these results are meaningful because recent work suggests that generative systems can exhibit systematic cultural tendencies and can shift outputs based on prompt framing and language conditions (Lu et al. 2025). In applied contexts, this makes cultural responsiveness a design problem. We do not simply get localization for free, and we may need explicit constraints, evaluation rubrics, or culturally responsive prompting strategies to avoid stereotyping and to ensure functional adequacy (Naidoo and Chadha 2025). The findings, stable genre skeleton with context-sensitive rhetorical strategies, fit this picture as the model preserves genre conventions, while cultural prompting nudges interpersonal tone and prioritization choices.

5.4 Authenticity, Destination Image, and CSR Legitimacy: How the Outputs Make Place and Make Trust

This section interprets the findings through three closely related lenses, (a) authenticity work in tourism discourse, (b) destination image/brand meaning, and (c) CSR legitimacy signaling. Across the corpus, the model repeatedly constructs meaningful travel by combining *place* (the Atlas), *guides/communities* (local grounding), and *responsible practice* (leave-no-trace, waste reduction, partnerships). What varies by target framing is less the presence of these resources than the emphasis and sequencing with which they are deployed.

Authenticity as staged access versus lived meaning. A long-standing point in tourism studies is that authenticity is not simply a property of a destination, but something produced in discourse and practice. In classic terms, tourism often promises access to a real behind-the-scenes culture (MacCannell 1976), while later work differentiates authenticity types, including experiential or existential modes tied to self-feeling and meaning rather than object verification (Wang 1999). The outputs largely avoid explicit exotic spectacle language; instead, they perform authenticity through local guide authority, community proximity, and ethical posture, responsible travel as a marker of realness. This aligns with the idea that tourism language can construct credibility and desirability by positioning the traveler within a particular moral and experiential frame (Dann 1996).

The observed Morocco-targeted tendency to foreground relational and exchange language, as shown in "people who know it well" and "genuine exchange", can be interpreted as a stronger existential/relational authenticity cue as connection, respect, and shared presence rather than solely as access to hidden places. That does not mean one audience values authenticity more; it means the model, under Morocco-target framing, more often selects authenticity resources that emphasize relationship and social closeness (Wang 1999).

The tourist gaze and the disciplined promise: safety, clarity, and ethical looking. Tourism discourse also works by shaping what is worth seeing and how it should be seen, and this is what Urry (2002) theorizes as the tourist gaze. In the corpus, the gaze is repeatedly disciplined. The traveler is invited to see the Atlas, but through a frame of controlled access (small groups, planning, trained guides) and moral constraint (respect, leave-no-trace). This matters because it suggests a hybrid promise; in other words, not just look at beautiful nature, but look correctly, i.e., in a way that is safe, respectful, and ethically legitimate (Dann 1996; Urry 2002). This hybrid promise is particularly visible when responsibility language is moved earlier in the sequence (or embedded implicitly), so that ethical travel becomes part of the destination hook, not only a CSR add-on.

Destination image and branding: from attributes to a coherent identity script. Destination-image models distinguish between cognitive beliefs, that is what attributes a place has, affective evaluations, meaning how it feels, and show how image formation links to intention and choice (Baloglu and McCleary 1999). The outputs consistently build a compact cognitive–affective package namely, (a) attributes ("2–5 days", "small group size", "logistics handled"), (b) affect ("respect," "confidence," "comfort"), and (c) an identity script ("eco," "thoughtful," "responsible"). This resembles the movement from place features to place meaning described in destination branding work that treats brands as organized narratives and associations, not just slogans (Pike 2002).

At the same time, the corpus shows brand meaning being anchored not only in the destination "High Atlas" but in the operator's ethos and governance. This aligns with branding perspectives that emphasize the managed identity of place-related organizations and the role of communication in building legitimacy and distinctiveness (Kavaratzis 2004). In short, the outputs do not rely solely on scenic descriptors; they repeatedly brand the experience through competence + responsibility + local grounding, which is consistent with tourism branding arguments that differentiation depends on more than physical attributes (Echtner and Ritchie 1993; Pike 2002).

CSR legitimacy: practice-heavy ethics versus explicit accountability. CSR communication does two things at once, it signals moral alignment (values) and it must also appear credible (evidence, specificity, accountability). The CSR outputs are strongest on *practice-based legitimacy*, they repeatedly describe responsible practices and partnerships as proof of ethical commitment. This is rhetorically powerful because it ties legitimacy to operational behavior rather than abstract claims (Dann 1996). However, the move-level finding that one CSR text omits a distinct accountability/reporting move (C4 absent) highlights a key tension. Legitimacy can be performed through values and practices, but readers may still look for explicit accountability cues when CSR is staged as a dedicated genre slot.

This tension also matters in AI-assisted marketing contexts. When organizations rely on generative systems, there is a risk of producing ethical polish text that sounds responsible but is under-specified on accountability. That risk is consistent with broader marketing/AI discussions that emphasize governance, transparency, and strategic control over automated communication (Huang and Rust 2018; Kietzmann et al. 2018; Kshetri et al. 2023). In other words, the CSR findings point to a practical implication, that is, responsibility language is easy for the model to generate, but accountability staging may require more explicit prompting constraints or editorial checks.

5.5 *Practical Implications for GenAI-Assisted Localization in Tourism SMEs*

The findings imply that GenAI localization work is best approached as a controlled production workflow rather than automatic translation and stylistic polish. In practice, the model reliably maintains a genre-faithful skeleton (HERO/ABOUT/CSR move repertoires), while target-framing prompts mainly shift tone, direct address, explicitness, and sequencing as rhetorical strategy. This has several concrete implications for practitioners designing, deploying, and evaluating AI-assisted tourism copy.

Treat GenAI as an augmentation system: managing tasks, not jobs. AI's most realistic organizational impact is often task-level substitution/augmentation, not full replacement of marketing work. A task-based view helps tourism SMEs decide which parts of copy production can be automated and which should remain human-led. This aligns with service-AI arguments that automation progresses from lower- to higher-level intelligence, with a prolonged augmentation stage before replacement is even plausible (Huang and Rust 2018). In the context, GenAI can efficiently generate variants, but humans must still verify accuracy, ethical claims, and cultural appropriateness.

Use genre-scaffold prompting to lock the skeleton, then localize the surface. Because the results show strong structural convergence by genre slot, practitioners should explicitly scaffold prompts by move sequence and only then ask for local adaptation of interpersonal and cultural cues. This reduces random drift and improves comparability across target versions.

From an advertising workflow perspective, this also fits how AI can support different stages of the customer journey by generating targeted messages while keeping a coherent path from awareness → evaluation → action (Kietzmann et al. 2018). Concretely, the first step named structure requires the model to output a move-labelled draft. The second step, named localization should specify which surface parameters may vary. The third step named post-edit should remove labels and polish into final web-ready copy.

This skeleton-first strategy is especially useful for SMEs because it produces repeatable outputs that can be reviewed quickly and revised systematically.

Build governance checks to avoid ethical polish in CSR. The CSR findings show that GenAI readily produces practice-oriented responsibility language, but accountability/reporting cues may appear in an inconsistent manner, for example when "C4 move" can be absent. For organizations, the risk is CSR that *sounds ethical* while remaining under-specified, particularly problematic where claims imply measurable impact.

This matters because GenAI marketing research flags both opportunities that have to do with speed, personalization, and ideation as well as challenges misleading content, governance gaps, and reputational risk in organizational use of generative systems (Kshetri et al. 2023). A practical remedy is to implement a CSR accountability checklist as a required gate before publication:

Does CSR include at least one measurable commitment (even minimal)?
Is there a clear monitoring/reporting cue (how you know it's working)?
Are partnerships/community claims stated in a way that can be verified?
Are leave-no-trace/sustainability claims framed with specific actions rather than generic virtue language?

For culturally responsive systems more broadly, governance and accountability should be treated as a design domain, not an afterthought, explicitly emphasized in CRAIF-C's lifecycle approach (Naidoo and Chadha 2025). In The workflow, Appendix A–style audits can be repurposed as a lightweight governance artefact. They document which expected moves/cues are present, implicit, or missing, and why.

Evaluate localization across languages and contexts, not just one English version. A key implication from recent evidence is that generative models can exhibit systematic cultural tendencies that shift with language conditions, and these shifts can influence downstream choices, including advertising recommendations (Lu et al. 2025). That means tourism SMEs should not assume that an English-only test guarantees behavior in Arabic/French or vice versa. If the organization operates multilingual pages, evaluation should include:

Parallel prompting across languages, holding intent constant,
Cross-language consistency checks for move structure and CSR accountability,
Human review by culturally knowledgeable readers for stereotyping, tone mismatch, or overclaiming.

CRAIF-C similarly stresses multi-context empirical testing and measurement tools as part of cultural responsiveness, alongside transparency and governance (Naidoo and Chadha 2025).

Use persona and audience framing carefully, guide it, don't stereotype it. The findings suggest that Morocco-targeted and USA-targeted affects stance/priority rather than genre identity. This supports a practical approach to persona design to guide the model using explicit audience needs and constraints, for example the use of "first-time hikers," "risk-averse travelers," "family groups," and "eco-conscious travelers", rather than relying on broad national stereotypes.

Where cultural dimension framing is used, it should be operationalized as optional, testable prompting constraints, not as fixed assumptions about people. Recent work on AI-assisted persona generation shows that cultural-dimension cues can be used as a structured guide but should be evaluated empirically for quality and appropriateness (Avis et al. 2025). In the setting, that translates into: "Try two persona framings, compare outputs with the rubric, and keep the one that improves clarity, trust, and non-stereotyped tone."

5.6 Managerial Urgency and Cultural Risk: Why Stable Skeleton Is Not Enough

The findings matter beyond genre theory because they speak directly to how organizations are already trying to scale customer-facing communication with generative AI. In marketing and advertising, AI is framed to transform large volumes of consumer-related signals into actionable insight across the consumer journey but this comes with recognized risk and governance concerns (Kietzmann et al. 2018). In that context, the key pattern, stable genre staging and variation mainly in rhetorical strategy, suggests a practical implication which is, quality control cannot stop at "does the copy contain the expected homepage moves?" It must also evaluate *how* those moves are realized, especially in openings, reassurance language, calls-to-action, and accountability/ethics cues.

This risk becomes more salient once cultural non-neutrality is treated as an empirical property of model behavior rather than a theoretical possibility. Prior evidence shows that generative models can exhibit systematic cultural tendencies across language conditions and that prompt framing can shift downstream recommendations and tone (Kshetri et al. 2023). Related work on culturally responsive AI also cautions that unmanaged cultural mismatch can erode trust and engagement, particularly when systems are deployed across diverse markets (Naidoo and Chadha 2025).

In the present analysis, move-presence patterns indicate relatively stable genre staging under controlled prompting, while the most consequential variation concentrates in interpersonal positioning and legitimacy-building strategies. For GenAI-assisted localization, this implies that governance should evaluate not only whether canonical moves appear, but how they are rhetorically realized because reputational and trust outcomes are more likely to be shaped by stance and legitimacy cues than by the mere presence of a complete move set. Taken together, these implications reinforce the study's central claim that genre staging is relatively stable under the controlled prompting, but cross-context risk concentrates in rhetorical realization and legitimacy cues.

6 Conclusions

This paper examined how GenAI-produced tourism website copy realizes genre-specific move structures and rhetorical strategies under controlled prompting across two prompt-defined target framings (USA-targeted vs. Morocco-targeted), using a multi-pass coding workflow with an explicit audit trail. Across HERO, ABOUT, and CSR, the outputs largely preserved genre-faithful move repertoires, supporting a stable skeleton and variable surface patterns. Variation concentrated less on whether expected communicative functions appeared and more on how they were realized, particularly through stance layering and through culture-linked cue families such as risk governance, explicitness signaling, and relational positioning. Interpreted cautiously, these contrasts are treated as tendencies in model outputs under audience-framing prompts rather than as claims about national communication behavior.

Substantively, the findings highlight that GenAI-assisted localization cannot be evaluated only at the level of move presence. The primary managerial risk lies in rhetorical realization namely, tone, direct address, legitimacy cues, and accountability staging, especially in CSR discourse where ethical polish can be produced without sufficiently explicit monitoring or reporting signals. Methodologically, the study contributes a replicable genre-sensitive procedure for analyzing GenAI outputs. Future work should test additional markets and languages, compare multiple models and timepoints, and include human evaluation of perceived trust, authenticity, and legitimacy to triangulate discourse patterns with reception outcomes.

7 AI-Assisted Writing Disclosure

Throughout the process of writing this paper, AI tools such as ChatGPT and Grammarly were used in a supportive manner specifically for language polishing, including grammar, spelling, punctuation, and refinements to clarity and academic style. In addition, Scribbr citation generator was used as an assisting tool to help with reference formatting and accuracy checks with the authors' careful verification. These tools were only used to enhance readability while preserving the authors' intended meaning; they were not used to generate the study's core content. Every modification was carefully reviewed and incorporated at the authors' discretion. The authors remain fully responsible for the final work and its integrity.

Coding Framework and Move-Presence Audit (Condensed)

This appendix summarizes the move labels used in Pass 1 (Table 1) and illustrates the Explicit/Implicit/Absent coding used in the move-presence audit matrix (Table 2).

Table 1 Move labels and functional definitions

Genre	Move	Functional definition
HERO	H1	Opening hook + positioning (what this is/for whom)
HERO	H2	Core offer + practical reassurance (trip basics + competence)
HERO	H3	Responsibility practices (ethics/sustainability actions)
HERO	H4	Closing payoff (final promise/outcome for the reader)
ABOUT	A1	Brand identity/origin/core values (who we are)
ABOUT	A2	Offer overview + audience positioning (what we do + for whom)
ABOUT	A3	Operational credibility (planning, safety, transparency, logistics)
ABOUT	A4	Responsibility/partnerships/sustainability practices (what we do responsibly)
ABOUT	A5	Closing promise/invitation (final takeaway)
CSR	C1	CSR stance/principle (what we believe).
CSR	C2	Stakeholders/partnership approach (who we work with; fairness)
CSR	C3	Environmental practices (impact reduction actions)
CSR	C4	Transparency/accountability (how we explain, review, improve)
CSR	C5	Closing goal/long-term impact (future-facing promise)

Table 2 Excerpt from move-presence audit matrix (illustrative cases)

Text_ID	Expected move	Presence	Notes
MOR-HERO-2	H4 (closing payoff)	Implicit	–
MOR-HERO-3	H3 (responsibility practices)	Implicit	–
MOR-ABOUT-1	A5 (closing promise/invitation)	Absent	A5 (invitation/CTA) absent: ABOUT copy ends on sustainability/values and operational reassurance without a booking prompt
MOR-CSR-2	C4 (transparency/ accountability)	Absent	C4 absent: No reporting/monitoring/ metrics/feedback mentioned; focus stays on practices + values

Legend (move-presence audit): Explicit = realized as a standalone segment; Implicit = function realized within another segment; Absent = not realized; NA = not applicable to the genre (Table 2).

References

Askehave, I., Nielsen, A.E.: Digital genres: a challenge to traditional genre theory. Inf. Technol. People. **18**(2), 120–141 (2005). https://doi.org/10.1108/09593840510601504

Avis, G.R., Nocco, S., Zhao, A.Q., Cremaschi, M.: Guiding AI-assisted persona generation with Hofstede's cultural dimensions: a mixed-methods evaluation. Behav. Inf. Technol., 1–29 (2025). https://doi.org/10.1080/0144929x.2025.2581258

Baloglu, S., McCleary, K.W.: A model of destination image formation. Ann. Tour. Res. **26**(4), 868–897 (1999). https://doi.org/10.1016/S0160-7383(99)00030-4

Bhatia, V.K.: Analysing Genre: Language Use in Professional Settings. Longman, London (1993)

Blain, C., Levy, S.E., Ritchie, J.R.B.: Destination branding: insights and practices from destination management organizations. J. Travel Res. **43**(4), 328–338 (2005). https://doi.org/10.1177/0047287505274646

Braun, V., Clarke, V.: Using thematic analysis in psychology. Qual. Res. Psychol. **3**(2), 77–101 (2006). https://doi.org/10.1191/1478088706qp063oa

Cyr, D.: Modeling web site design across cultures: relationships to trust, satisfaction, and e-loyalty. J. Manag. Inf. Syst. **24**(4), 47–72 (2008). https://doi.org/10.2753/MIS0742-1222240402

Dann, G.M.S.: The Language of Tourism: a Sociolinguistic Perspective. CAB International, Oxford (1996)

De Mooij, M., Hofstede, G.: The Hofstede model: applications to global branding and advertising strategy and research. Int. J. Advert. **29**(1), 85–110 (2010). https://doi.org/10.2501/S026504870920104X

Diaz, D., Stöckl, A.: Cultural differences in perception and engagement of AI-generated online ads. In: Human Interaction and Emerging Technologies (IHIET-AI 2025), vol. 161, pp. 252–259 (2025) https://doi.org/10.54941/ahfe1005916

Echtner, C.M., Ritchie, J.B.: The measurement of destination image: an empirical assessment. J. Travel Res. **31**(4), 3–13 (1993). https://doi.org/10.1177/004728759303100402

Gevorgyan, G., Manucharova, N.: Does culturally adapted online communication work? A study of American and Chinese internet users' attitudes and preferences toward culturally customized web design elements. J. Comput.-Mediat. Commun. **14**(2), 393–413 (2009). https://doi.org/10.1111/j.1083-6101.2009.01446.x

Hall, E.T.: Beyond Culture. Anchor Press/Doubleday, New York (1976)

Hofstede, G.: Culture's Consequences: Comparing Values, Behaviors, Institutions and Organizations across Nations, 2nd edn. Sage Publications, Thousand Oaks, CA (2001)

House, R.J., Hanges, P.J., Javidan, M., Dorfman, P.W., Gupta, V. (eds.): Culture, Leadership, and Organizations: the GLOBE Study of 62 Societies. Sage Publications, Thousand Oaks, CA (2004)

Huang, M., Rust, R.T.: Artificial intelligence in service. J. Serv. Res. **21**(2), 155–172 (2018). https://doi.org/10.1177/1094670517752459

Kavaratzis, M.: From city marketing to city branding: towards a theoretical framework for developing city brands. Place Branding. **1**(1), 58–73 (2004). https://doi.org/10.1057/palgrave.pb.5990005

Kietzmann, J., Paschen, J., Treen, E.: Artificial intelligence in advertising: how marketers can leverage artificial intelligence along the consumer journey. J. Advert. Res. **58**(3), 263–267 (2018). https://doi.org/10.2501/JAR-2018-035

Kshetri, N., Dwivedi, Y.K., Davenport, T.H., Panteli, N.: Generative artificial intelligence in marketing: applications, opportunities, challenges, and research agenda. Int. J. Inf. Manag. **75**, 102716 (2023). https://doi.org/10.1016/j.ijinfomgt.2023.102716

Lu, J.G., Song, L.L., Zhang, L.D.: Cultural tendencies in generative AI. Nat. Hum. Behav. **9**(11), 2360–2369 (2025). https://doi.org/10.1038/s41562-025-02242-1

MacCannell, D.: The Tourist: a New Theory of the Leisure Class. Schocken Books, New York (1976)

McSweeney, B.: Hofstede's model of national cultural differences and their consequences: a triumph of faith—a failure of analysis. Hum. Relat. **55**(1), 89–118 (2002). https://doi.org/10.1177/0018726702551004

Naidoo, V., Chadha, K.K.: Culturally responsive AI chatbots: from framework to field evidence. Comput. Hum. Behav. Artif. Hum.S. **6**, 100224 (2025). https://doi.org/10.1016/j.chbah.2025.100224

Pike, S.: Destination image analysis— a review of 142 papers from 1973 to 2000. Tour. Manag. **23**(5), 541–549 (2002). https://doi.org/10.1016/S0261-5177(02)00005-5

Singh, N., Zhao, H., Hu, X.: Cultural adaptation on the web: a study of American companies' domestic and Chinese websites. J. Glob. Inf. Manag. **11**(3), 63–80 (2003). https://doi.org/10.4018/jgim.2003070104

Singh, N., Zhao, H., Hu, X.: Analyzing the cultural content of web sites: a cross-national comparison of China, India, Japan, and US. Int. Mark. Rev. **22**(2), 129–146 (2005). https://doi.org/10.1108/02651330510593241

Singh, N., Pereira, A.: The Culturally Customized Web Site: Customizing Web Sites for the Global Marketplace. Elsevier Butterworth-Heinemann, Oxford (2005)

Swales, J.M.: Genre Analysis: English in Academic and Research Settings. Cambridge University Press, Cambridge (1990)

Trompenaars, F., Hampden-Turner, C.: Riding the Waves of Culture: Understanding Cultural Diversity in Business, 2nd edn. Nicholas Brealey Publishing, London (1997)

Urry, J.: The Tourist Gaze, 2nd edn. Sage Publications, London (2002)

Wang, N.: Rethinking authenticity in tourism experience. Ann. Tour. Res. **26**(2), 349–370 (1999). https://doi.org/10.1016/S0160-7383(98)00103-0

Wodak, R., Meyer, M. (eds.): Methods of Critical Discourse Analysis, 2nd edn. Sage Publications, London (2009)

Mapping the Research Landscape of AI Startups: A Bibliometric Analysis of Incubation and Acceleration Ecosystems (2011–2025)

Walid Dalal, Dounia Rabhi, and Marouane Ikira

Abstract This research analyzes the literature around the convergence of artificial intelligence startups and entrepreneurial support systems (such as incubators and accelerators). As the importance of both AI-driven entrepreneurship and incubation models has risen in recent years, the body of scientific work that integrates these two areas is very limited. This study identifies the relationship between these two fields by utilizing a bibliometric methodology on a sample set of 430 peer-reviewed articles from 2011 to 2025 that are indexed in the Web of Science Core Collection. The study generated descriptive indicators and visualized keywords, thematic clusters and the intellectual structure of the literature using Bibliometrix (R/Biblioshiny) and Vos viewers. The results indicate a growing and consolidating area of research characterized by a movement away from traditional incubation and innovation themes towards digital transformation, venture finance and AI-enabled entrepreneurial ecosystems. The findings further demonstrate how incubators and accelerators are evolving into data-driven, networked and value-added platforms that integrate fintech, predictive analytics and algorithmic support. Additionally, this study uncovers the leading authors, institutions and countries that are contributing to this body of literature and identifies many gaps within the literature in relation to the socio-institutional and ethical implications of AI integration in startup support systems.

Keywords AI start-ups · Business incubators and accelerators · Entrepreneurial ecosystems · Digital transformation · Bibliometric analysis

W. Dalal (✉) · D. Rabhi · M. Ikira
Faculty of Legal, Economic, and Social Sciences, LARGESS, Chouaib Doukkali University, El Jadida, Morocco
e-mail: dalal.w660@ucd.ac.ma

D. Rabhi
e-mail: rabhi.d@ucd.ac.ma

M. Ikira
e-mail: Ikira.m@ucd.ac.ma

A. Bouziane and K. Bouziane (eds.), *AI, Management, and Culture: Shaping Organizations in the Digital Age*, Studies in Computational Intelligence 1291,
https://doi.org/10.1007/978-3-032-27882-1_8

1 Introduction

The last decade has seen an unprecedented convergence of the areas of entrepreneurship, digital transformation, and artificial intelligence (AI) occurring in the arena of Global Innovation, with Startups being proactive engines of technological, economic, and social change by producing data-driven products and platforms (Eesley and Miller 2018). In this new environment, Incubators and Accelerators have emerged as critical intermediaries between early-stage ventures and the resources, mentoring, and legitimization they need to thrive in rapidly changing marketplaces (Cohen and Hochberg 2014; Mian 2014). The functions of these mechanisms are no longer limited to providing space or mentoring programs as they have evolved into strategic ecosystem participants combining AI, fintech and the creation of digital infrastructures as mechanisms for supporting high-growth Technology Ventures (Mohammadi and Sakhteh 2023; Petrucci 2018).

While the body of academic research on the design, operation, and effects of incubators and accelerators on entrepreneurs is increasing (Brun 2019; Pauwels et al. 2016; Stayton and Mangematin 2019), the influence of AI and advanced digital technologies on the efficacy of these mechanisms has not yet been thoroughly examined. Most studies examining the various organizational and managerial aspects of Incubator programs (Hackett and Dilts 2004; Mian 2014) or the financial and policy structures surrounding entrepreneurial activities (Grimaldi and Grandi 2005; Samuelsson and Jutterström 2024) do not address how the introduction of AI-driven technologies are changing the dynamics of incubators, accelerators, and methods of startup financing from Venture Evaluation and Investment Decisions to Mentoring and Network Orchestrating. Because of the fragmentation of research in this area, we do not currently have a comprehensive understanding of how AI and other emerging technologies interact with the institutional and social aspects of Entrepreneurial Ecosystems.

As innovation systems around the world begin to more heavily rely on data-based evaluation and algorithmic decision-making when identifying and funding potentially successful ventures, the gap identified above is of particular importance. The increasing reliance on technological advancements (such as AI) has seen the development of more sophisticated incubators and accelerators that use predictive analytics, digital dashboards, and algorithmic matching to facilitate investment flows and allocate resources more effectively (Firdaus et al. 2019; Eesley and Miller 2018). However, in spite of these advances in technology, scientific research mapping the connections between AI businesses, incubation and acceleration remains scarce. As a result, a systematic mapping of this intersection should provide insight into important convergences of thought, identifying blind spots for empirical research, and revealing the emerging trends in this rapidly evolving area of study.

This study will seek to fill this gap through the development of a comprehensive bibliometric analysis of publications published between 2011 and 2025 to map out the way that the literature describes how incubation and acceleration literature has developed in relation to the fields of AI and digital entrepreneurship. By analyzing all of the data contained in English-language 430 articles in the Web of Science Core Collection, this study will be able to provide an explanation of how the intellectual structures, thematic developments and collaboration networks are working within this area of research.

Accordingly, the central research question guiding this study is:

How has the academic understanding of incubation and acceleration evolved in relation to AI startups and digital innovation between 2011 and 2025?

To operationalize this question, three sub-questions are proposed:

1. What are the main thematic and conceptual trends in the literature on incubation, acceleration, and AI-driven entrepreneurship?
2. Which authors, institutions, and countries have been most influential in shaping this interdisciplinary research landscape?
3. How have research themes evolved over time, and what are the emerging or declining areas of inquiry?

The major knowledge gap surrounding traditional incubation research and the emerging literature related to AI entrepreneurs and fintech-enabled innovation ecosystems will be at least partially filled by answering the questions identified in this study. This research adds theoretical and methodological contributions, with the theoretical contributions illustrating how incubators and accelerators are defined in a technology-driven world, influenced by the use of automated systems, data and collaborative networks (Spigel 2017; Petrucci 2018). The methodological contribution of this research provides a baseline mapping of global research trends in this specific area of inquiry through the application of bibliometric analysis tools (e.g. Bibliometrix, VOSviewer) that effectively visualize how the intellectual landscape of this area of scholarship has evolved over time.

Ultimately, scholars, policymakers and practitioners who wish to design effective innovation incubators must take time to learn about the relationships between AI technology, incubators, and startup incubators' ecosytems. As digital transformation continues to change the face of entrepreneur finance, funding and support systems, researchers must first define the body of literature that exists on the subject in order to begin developing frameworks that would be adaptive, equitable and data-driven for the future of AI-empowered entrepreneurs.

The remaining portions of this paper will include: Sect. 2 provides the literature background, Sect. 3 explains the bibliometric methodology, Sect. 4 presents the results, Sect. 5 discusses the main implications for research and practice, and Sect. 6 concludes the paper.

2 Literature Review

The previous findings of incubation and acceleration research, in conjunction with the ongoing development of entrepreneurial support systems during our era of technological advancement, have evolved over time to reflect the advancement of AI-enhanced entrepreneurship and the establishment of digital incubation communities. Incubators have served as vehicles for creating new businesses, creating regions where businesses can thrive, and supporting entrepreneurial endeavors through all levels of business (Hackett and Dilts 2004).

The initial phases of incubation research established that incubators were a process by which entrepreneurs are drawn to each other in order to interact with each other during the entrepreneurial journey and learn from one another to develop better businesses. Incubators provide a structured environment for entrepreneurs to work together; therefore, by utilizing structured resource support, (1) an increase in innovation can occur, (2) new businesses are created in cities such as Denver and Kansas City; and (3) startup failures are reduced through the use of structured incubators. Grimaldi and Grandi (2005) developed a classification of incubator models as "business" incubators (meaning that they are solely focused on growing businesses), "university" incubators (which are established and maintained by universities), and "hybrid" incubators (those that utilize both the models described above for supporting entrepreneurship). Grimaldi and Grandi (2005) are important to the understanding of how incubators can create trade-offs between the goals of public policy and the success of entrepreneurs. In the maturation of knowledge about incubators, researchers have been able to develop frameworks for the understanding of the relationship of incubators to entrepreneurs.

Mian (2014) synthesized decades of incubation research, stating that incubators have transitioned from being a physical resource support (the buildings provide the actual physical location, the equipment to provide services, etc.) to being a critical hub of the exchange of knowledge, technology transfer, and commercial development of ideas, products, and services within innovation ecosystem networks. Simultaneously, Spigel (2017) has presented a conceptualization of entrepreneurial ecosystems that represent the relationships between entrepreneurs and their collaborative entities in society, and how all the social, cultural, and material factors support or constrain entrepreneurship at the local level. This shift focuses on understanding how incubators, accelerators, and regional policies that affect innovation create entrepreneurial growth has created additional opportunities for research and intelligence.

As the concept of 'acceleration' has developed, there has been an opportunity for a distinct type of organization that follows the accelerator model. Cohen and Hochberg (2014) were the first to create an organized grouping for accelerators indicating that they are 'programs designed to accelerate the development of new ventures through the provision of funding, mentoring, and cohort-based learning, thereby compressing the timeline of venture development.' Other researchers have defined accelerators in relation to venture capital and entrepreneurial education suggesting that accelerators have become 'investment-driven platforms that blur the line between venture capital and entrepreneurial education.' Pauwels et al. (2016) added to our understanding of accelerator programs by providing a way to classify accelerator configurations based on their strategic objectives (ecosystem builders, corporate and public accelerators) which acknowledges the various types of institutional logics influencing how startups are accelerated. Together, these authors frame accelerators as value creating systems rather than only as a source of funding or training.

Recent research has modified this approach by applying value chain and relational models of incubation. Using Brun's (2019) analogy of incubators being 'start-up factories,' Mohammadi and Sakhteh (2023) provided an example of how these incubators/accelerators create value through their process efficiency and outcome measurement. By completing a systematic review of the literature, they built a framework for understanding accelerators/accelerator programs via a multi-staged approach to value creation (selection, support, and scaling). In doing so, they were able to show how accelerators institutionalize the venture creation process and use the framework to incorporate evaluation and funding into defined phases of the process in multiple ways. Likewise, Petrucci (2018) further developed our understanding of incubation processes as relational rather than linear, showing that the success of incubated ventures is a function of the ability to integrate them into the dynamic nature of business networks rather than by programming or operationalizing those ventures within a linear sequence.

The development of incubation strategies has led to greater scrutiny of the effectiveness and accessibility of these programs through research. According to Lukosiute et al. (2019), participation in accelerators or incubators does not necessarily lead to improved results; rather, post-incubation growth may be influenced by the potential for dependency and misalignment of release strategies. Amelia et al. (2021) conducted a similar analysis of government-funded incubators in Indonesia and found that while these programs promote inclusivity, they often suffer from weaknesses related to scalability and the depth of innovation.

Hackett and Dilts (2004) and Grimaldi and Grandi (2005) were among the first to critique their effect of context and management on the success of entrepreneurship incubation versus standardization; as a result, they expected that successful incubation would depend upon context and management, NOT on the same level of support. A new stream of research is exploring digital and artificial intelligence based transformation within incubating ecosystems. Firdaus et al. (2019) found that many incubators have begun to use mobile applications and artificial intelligence notification systems in order to support communication and monitoring of entrepreneurs.

Mohammadi and Sakhteh (2023) identify the increasing reliance on data-driven evaluation frameworks to monitor and predict whether or not funding will be allocated to startups during the accelerator program. Eesley and Miller (2018) provide a more comprehensive picture of the impact that data and digital technology has had on the way that universities and innovation centers produce ambitious entrepreneurs, particularly in the form of data-based evaluations and networks of funders. This convergence of digital, AI and entrepreneurial technologies is indicative of a fundamental shift towards data-driven and technology-supported incubation.

Several studies have shown how important social and institutional support can be for developing effective incubation programmes. Samuelsson and Jutterström (2024) looked at a sponsorship model, and found that building legitimacy through mentoring, as well as building strong relationships, was far more crucial to start-up performance than creating facilities. Petrucci (2018) reviewed a number of mid-stage incubators and found that building trust and having adaptive forms of governance were necessary for a system to be able continuously to support start-ups in achieving success. Finally, Mian (2014) showed that the level of stakeholder alignment with regards to the maturity of ecosystem had critical importance for developing sustainable incubation programs through links between institutions, governments, and investors. Collectively, these studies indicate that the success of an incubator requires the incorporation of all three input factors: digital assets, placement of financial resources, and establishment of effective governance frameworks supported by robust relationships. Although the vast majority of studies on incubation strategies were originally conducted in North America and Western Europe as the regions with the most developed innovation structures and the highest concentrations of venture capital and accelerator programs (Stayton and Mangematin, 2019), other emerging regions have begun to provide valuable contextual data. For example, Amelia et al. (2021) demonstrated how adaptive models in Indonesia are evolving, while Mohammadi and Sakhteh (2023) provided an explanation of the changes in the Iranian accelerator market, highlighting how the global diffusion of incubation has been accompanied by adaptive changes in local institutional contexts. The growing body of comparative studies demonstrates a converging set of principles of acceleration being established globally, particularly regarding the growing use of data and artificial intelligence, and increasing collaboration among countries, while at the same time the availability of resources and effective governance continues to show signs of regional divergence.

The literature consistently finds that incubators and accelerators have become more strategically driven and embedded in relationships and data and less focused solely on providing resources. The emergence of AI and fintech tools within incubators and accelerators represents another key milestone in the professionalization of start-up ecosystems; however, it will also prompt future studies to explore issues of access, transparency, and inclusivity based on the convergence of incubators, investors, and innovation policy. Future research should explore how digital tools such as AI and fintech will ultimately help balance efficiency and equity in incubation so that technological advancements lead to sustainable global innovation.

3 Methodology

3.1 Data Collection

The bibliometric dataset was compiled from the Web of Science Core Collection, chosen for its comprehensive coverage of peer-reviewed journals in management, entrepreneurship, and technology. Two sequential search queries were performed to ensure both breadth and specificity.

The first query targeted incubation and acceleration frameworks:

("incubator" OR "business incubator" OR "startup incubator" OR "accelerator" OR "business accelerator" OR "innovation hub" OR "venture builder" OR "entrepreneurship center" OR "startup program" OR "incubation program" OR "acceleration program").

The second query focused on startups and technology-driven ventures:

("startup" OR "start-up" OR "new venture" OR "entrepreneurial firm" OR "tech startup" OR "technology-based startup" OR "AI startup" OR "artificial intelligence startup" OR "machine learning venture").

The initial query retrieved 95,227 documents, which were narrowed down to 758 after applying the second search string. Filters were applied to include only articles (548), published between 2011 and 2025 (445), and written in English, resulting in a final dataset of 430 publications used for analysis.

3.2 Data Processing and Software Tools

The bibliographic information was imported in plain text format and then was imported into RStudio and analysed with Biblioshiny. The results include generating descriptive metrics (e.g., Citation patterns, Annual Growth Rate, Authorship pattern), and how the data's conceptual structure can be mapped to support hypothesis development. Additionally, network visualization of the data have also been generated, including co-occurrence of keywords, mapping of thematic typologies, where clustering of data points by co-occurrences.

3.3 Analysis Procedure

The analysis followed three main stages:

- Descriptive Bibliometric Analysis—assessing publication trends, sources, authors, affiliations, and country contributions to identify the structure of the research field.

- Thematic and Evolutionary Analysis—exploring the intellectual development of key concepts (e.g., innovation, incubation, accelerators) and their evolution toward topics such as AI financing, fintech, and digital transformation.
- Network Visualization and Clustering—using co-occurrence mapping to identify thematic clusters linking entrepreneurship, technology, and finance, thereby uncovering the conceptual and financial underpinnings of AI startup ecosystems.

The study covers the period 2011–2025, reflecting the maturation of research on AI startup financing and incubation ecosystems. This timeframe captures both early conceptual work on accelerators and more recent applied studies addressing AI-driven innovation and investment models (Table 1).

Table 1 Main informations

Description	Results
Main information about data	
Timespan	2011–2025
Sources (Journals, Books, etc.)	246
Documents	430
Annual growth rate %	− 8.01
Document average age	5.55
Average citations per document	18.28
References	18,548
Document contents	
Keywords plus (ID)	784
Author's keywords (DE)	1415
Authors	
Authors	1387
Authors of single-authored documents	68
Authors collaboration	
Single-authored documents	70
Co-authors per document	3.58
International co-authorships %	30.47
DOCUMENT TYPES	
Article	430

Source Made by authors using biblioshiny

4 Results

4.1 Descriptive Bibliometric Indicators

4.1.1 Main Informations

From the years 2011–2026, the research on artificial intelligence (AI) start-up/acceleration ecosystems was studied across 430 papers from 246 different sources; this indicates the development of a sound base for future research into these areas. The data analysis indicates that while research interest peaked in the early part of the last decade, there has been an overall decline (regression) since that time (−8.01% growth). This indicates that the research has matured into specialized and applied studies; and is therefore stabilizing rather than expanding. Average citation count (18.28/citation) is moderate, which means their influence remains moderate. Some have even been referred to as core references in the field of Innovation and Entrepreneurship. The publication age of papers is approximately 5.5 years and thus reflects current patterns and ongoing conversation about how emerging AI technologies will continue to evolve. The total reference count (18,548) and the number of descriptive author keywords (1415) show the interdisciplinary approach taken by scholars when researching this area, including aspects of Technology Management, Entrepreneurship, and Digital Transformation. Collaboration between scholars is substantial with an average of 3.6 authors on each paper, as well as 30.47% of papers being co-authored by international scholars, indicating that AI Start-Up Ecosystems are a globally oriented, cross-institutional engagement, a pattern consistent with the interconnected and innovation-driven nature of AI startup ecosystems (Table 2).

4.1.2 Annual Scientific Production

AI Startup & Incubation/Acceleration publishes continuously through 2011 and peaked in 2025 with 73 published works. 2026 represents early works published in 2025 and is expected to stabilize. The highest Average Citations per Document were in 2012 (87.6) and 2019 (39.83), illustrating how foundational studies from those years impacted later research. Research was strong between 2020–2021 but citation impact has decreased, as evidenced by an average MeanTCperYear of 2–3. This average reflects a growing body of work in the AI incubation ecosystem that is relatively new and is still accumulating prestige. The annual growth rate of − 8.01% demonstrates that this field of research is mature; it is transitioning from conceptualization (research) to more empirical and applied studies (Tables 3 and 4).

Table 2 Annual production and citation growth

Year	MeanTC per article	Documents (N)	MeanTC per year	Citable years
2011	16.29	7	1.02	16
2012	87.60	10	5.84	15
2013	29.75	16	2.12	14
2014	28.50	10	2.19	13
2015	22.90	10	1.91	12
2016	16.82	11	1.53	11
2017	29.16	19	2.92	10
2018	24.73	40	2.75	9
2019	39.83	42	4.98	8
2020	20.22	32	2.89	7
2021	18.64	39	3.11	6
2022	11.92	36	2.38	5
2023	8.59	37	2.15	4
2024	5.67	46	1.89	3
2025	1.26	73	0.63	2
2026	3.50	2	3.50	1

Source Made by authors using biblioshiny

Table 3 Most relevant sources

Source	Articles
Journal of technology transfer	16
Sustainability	15
Fusion engineering and design	8
IEEE transactions on engineering management	8
International journal of entrepreneurial behavior and research	8
Technological forecasting and social change	8
Technovation	8
Small business economics	7
European journal of innovation management	6
Journal of business venturing	6

Source Made by authors using biblioshiny

4.1.3 Most Relevant and Cited Sources

Journals bridging innovation, entrepreneurship, and technology management are at the center of the publication landscape. The Journal of Technology Transfer (16 articles) and Sustainability (15 articles) are two journals that have consistently produced

Table 4 Most cited sources

Source	Articles cited
Journal of business venturing	1017
Technovation	923
Research policy	679
Entrepreneurship theory and practice	610
Strategic management journal	546
Small business economics	527
Academy of management journal	493
Journal of technology transfer	469
Academy of management review	401
Organization science	349

Source Made by authors using biblioshiny

articles in this space, indicating a focus on how knowledge is diffused and on the framework for developing sustainable innovations. Citations support the dominance of the Journal of Business Venturing (1017 citations), Technovation (923 citations), and Research Policy (679 citations). This indicates that the theoretical context for the field is shaped by entrepreneurship literature and innovation policies. The patterns of publication and citation reflect a research area that includes elements of both applied and emerging scholarships in specialized academic journals and older, more established sources of conceptual depth (Tables 5, 6 and 7).

Table 5 Most relevant authors

Author	Articles	Articles fractionalized
Van Rijnsoever, Frank J	5	1.58
Banka, Michal	4	0.95
Cabezas, Paula	4	1.12
De-Pablos-Heredero, Carmen	4	0.85
Jimenez, Francisco Jose Blanco	4	0.90
Asensio-Ciria, Ana	3	0.65
Assenova, Valentina A	3	2.00
Avnimelech, Gil	3	1.25
Eveleens, Chris P	3	1.00
Kanbach, Dominik K	3	1.03

Source Made by authors using biblioshiny

Table 6 Most relevant affiliations

Affiliation	Articles
Universidad Rey Juan Carlos	16
University of Colorado system	12
University of Colorado Anschutz medical campus	10
Harvard university	9
University of London	9
Aarhus university	8
Italian national agency for new technical energy and sustainable economic development	8
Swiss federal institutes of technology domain	8
University of Massachusetts system	8
University of Tehran	8

Source Made by authors using biblioshiny

Table 7 Most relevant countries

Country	Articles	Articles %	SCP	MCP	MCP %
USA	64	14.9	51	13	20.3
China	26	6.0	23	3	11.5
United Kingdom	26	6.0	16	10	38.5
Italy	24	5.6	15	9	37.5
Germany	21	4.9	13	8	38.1
Canada	18	4.2	13	5	27.8
Spain	18	4.2	13	5	27.8
Netherlands	16	3.7	7	9	56.3
Russia	14	3.3	11	3	21.4
India	13	3.0	12	1	7.7

Source Made by authors using biblioshiny

4.1.4 Most Relevant Authors, Affiliations and Countries

Despite being globally connected with a high concentration of research, authorship and institutional patterns point toward a research landscape that consists of a number of different areas or themes. Van Rijnsoever, Frank, J. is at the top of the list with five publications, while Assenova (with 2.00 fractionalized) and Avnimelech (with 1.25) represent strong individual contributions to the topic of innovation and entrepreneurship, indicating that these individuals really have concentrated expertise in that area. The main institutional hubs for this region are the Universidad Rey Juan Carlos (with 16), the University of Colorado System (with 12), and the two foremost universities in the world, Harvard and ETH Zurich, are also key partners that allow for much

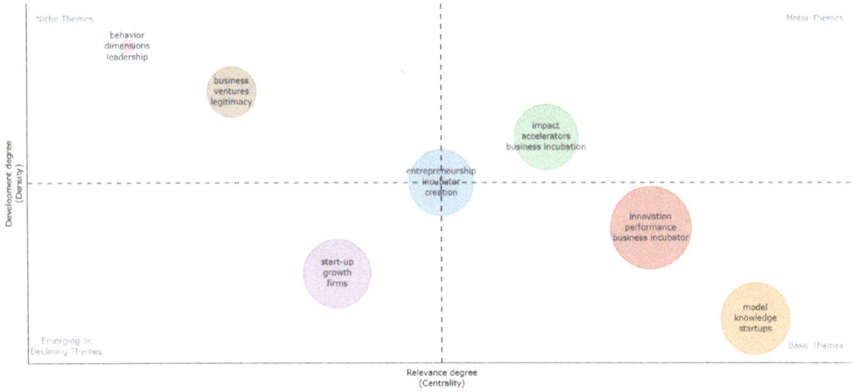

Fig. 1 Thematic map. *Source* Made by authors

balanced transatlantic collaboration between these two locations. While the USA accounts for a majority of research output (14.9%), the United Kingdom, Italy, and the Netherlands demonstrate very high rates of collaboration with other European nations, as high as 56.3%. This indicates that innovation and entrepreneurship is an area of research that has matured as a field, and that it is becoming more internationally connected through cooperation between both scholars and institutions in various parts of Europe and North America (Fig. 1).

4.2 Thematic Structure and Evolution

A thematic map demonstrating how financing for AI startups fits into overall incubation and innovation research. The "*entrepreneurship, incubator, creation*" clusters indicate that funding opportunities are closely tied to both venture formation and ecosystem development. Key themes like "*impact*" and "*accelerators*", along with "*innovation*" and "*performance*" relating to business incubators, indicate how financial support, accelerator models, and innovation result from existing research. Basic themes of "*model, knowledge, startups*" create the theoretical framework for understanding financing and knowledge transfer in ai ventures. Conversely, niche themes such as "*leadership*" and "*legitimacy*" focus on how investor trust/legitimacy impact a new venture's ability to secure funding, whereas the emergence of themes on "*start-up growth firms*" continues to highlight the focus on scaling and post-incubation financing challenges related to ai entrepreneurship (Fig. 2).

The thematic evolution demonstrates that researchers have shifted their focus away from broad subjects of entrepreneurship and innovation to studying the specific financial and technological aspects of how AI and fintech are currently being funded and marketed. From the years 2011–2015, the area of study for researchers was on performance (7 links) and innovation (6 links), both of which represent some

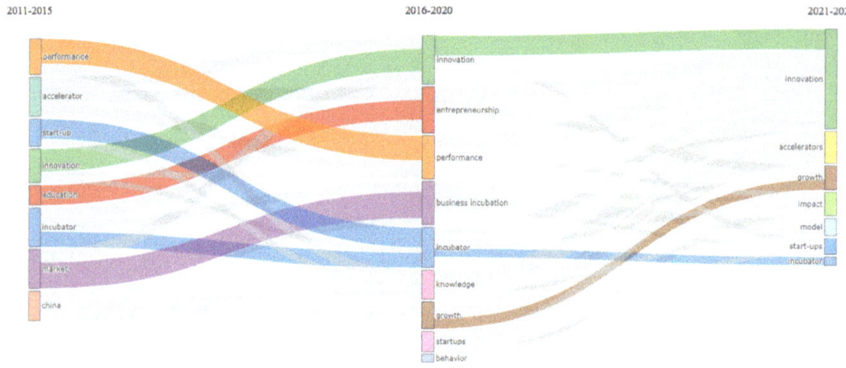

Fig. 2 Thematic evolution. *Source* Made by authors using biblioshiny

of the first attempts to understand how technology-based businesses could compete in the marketplace. Between 2016 and 2020, the concept of innovation (9 links) became the most important area of focus. At this time, many more connections were made between entrepreneurship and innovation (8 links for each), which reflects the increasing numbers of incubators and financial support systems established for both AI and digital businesses (more structure for support). Most recently (from 2021 to 2025), innovation continues to be the main area of focus but is now heavily related to the establishment of accelerators, growth, and impact in the ecosystem. This reflects the effect that these models and AI-driven innovations have had on startup funding, scalability, and their impact on the ecosystem. The continuous appearance of incubators and startups suggests that an institutional foundation has been established and that, consequently, new developments in the area of AI-enabled funding, investment analytics, and performance evaluation frameworks are being researched within tech entrepreneurship studies (Fig. 3).

The co-occurrence keyword network shows an intertwined research area that connects Entrepreneurship, Technology and Finance. These are linked through eight Thematic Clusters.

The first cluster map (11 items) defines the core clusters of the field: Innovation, Accelerator, Digital Transformation, FinTech and Growth combined with Public Policy. The results from this cluster indicate that more investment on AI and fintech is linked to policy frameworks that support technological innovation, illustrating how Digital Transformation influences the way that investors will invest in new venture models for their clients.

The second cluster (11 items) is concerned with Entrepreneurial Learning, Human Capital and Social Networks and identifies the specific methods and skills which are utilized to support AI ventures being Incubated. The third cluster (9 items) focuses on Companies, Ecosystems, Technology, Innovation and highlights both corporate participation and ecosystem orchestration as it pertains to Funding AI Startups.

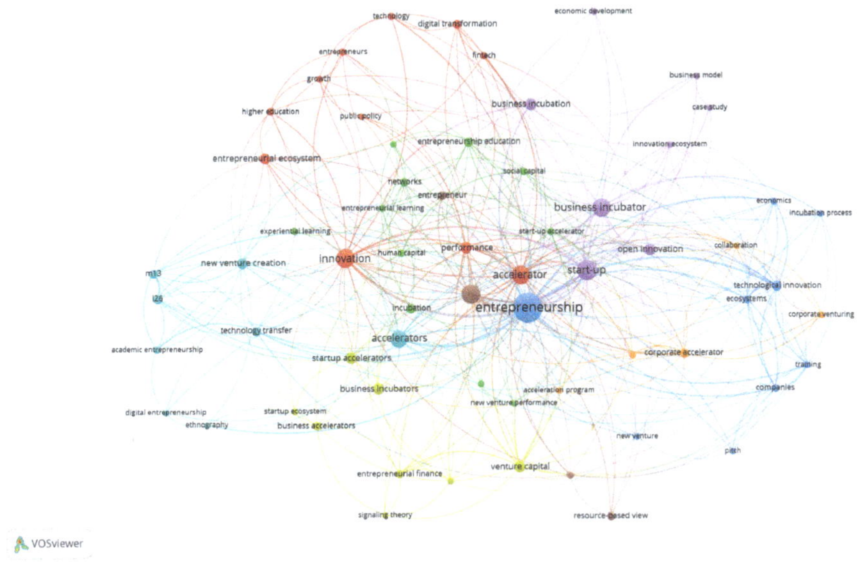

Fig. 3 Keywords co-occurrence (Min co-occurrence of 4). *Source* Made by authors using VosViewer

The fourth cluster (8 items) links specifically to Venture Capital and Business Accelerators to produce the Financial Backbone of the field of AI Entrepreneurship. And the fifth cluster (8 items) looks at the Business Incubator and Innovation Ecosystem and establishes the relationship between structured programs and transforming research into commercialized AI Technologies.

The sixth to the eighth clusters provide further insights: Cluster 6 addresses the relationship between digital and academic entrepreneurship; Cluster 7 relates to corporate acceleration and collaboration; And cluster 8 links commercialization with the resource-based view.

Overall, the map clearly illustrates that AI Startups require Financial Support embedded in a much broader context which includes Digital Transformation, Knowledge Transfer, as well as Venture Capital, thus demonstrating that Innovation Policy, Entrepreneurial Capability and Financial Infrastructure are all converging to impact the development of AI-driven startup ecosystems (Fig. 4).

The overlay visualization has demonstrated how the research fields of innovation, incubation and entrepreneurship have evolved to today's topics of digital transformation, fintech and corporate acceleration (highlighted in yellow, 2021–2023). As shown above with the overlay visualization, core themes related to entrepreneurship, innovation, accelerator and incubator continue to dominate the study of this field and demonstrate their ongoing importance as foundational concepts. A growing number of connections with venture capital, entrepreneurial finance, and technology transfer further illustrate an increased emphasis on funding mechanisms and integration of ecological systems for ai and technology startups. These new developments signal

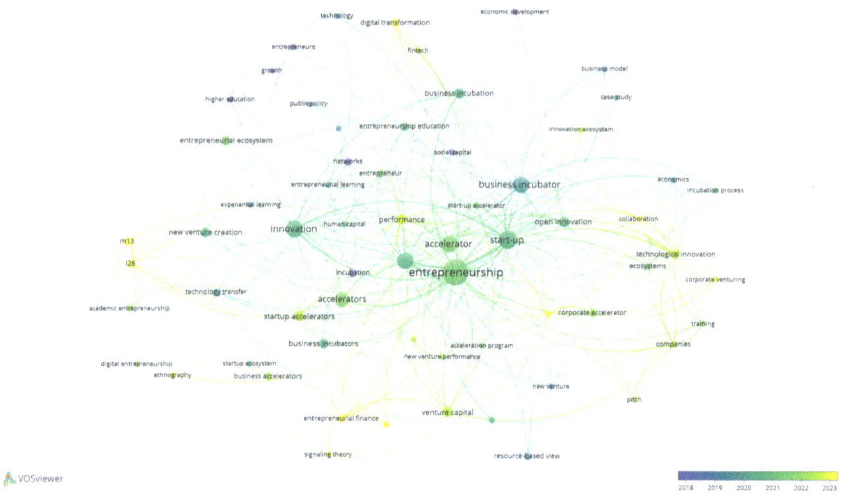

Fig. 4 Keywords overlay visualization. *Source* Made by authors using Vosviewer

a transition from primarily theoretical frameworks to more applied funding-based approaches to technology and include the growing trend of research in this area being driven by finance and technology.

5 Discussion

The bibliometric review clearly indicates that there is an evolution and therefore maturation of the Accelerator/Incubator and AI-based Innovation. As such, there is growing research and practitioner interest in how these programs or models currently support Entrepreneurs and how they will be managed in the future. Nevertheless, there are differing opinions regarding their effectiveness and inclusivity. The early literature identified accelerators and incubators (Stayton and Mangematin 2019) as vital facilitators of entrepreneurship through the provision of nets for resource networking, creating legitimacy, and accelerating the survival of new ventures. Alternatively, more recent studies have taken a more critical stance toward the limitations of context, differences in performance outcomes, and the complexity of digital transformation and AI integration (Mohammadi and Sakhteh 2023; Samuelsson and Jutterström 2024).

All studies have a common agreement that incubators and accelerators provide value for new ventures through mentorship, capital investment, and access to networks. These enrich the potential for new startups to grow and create opportunities. Brun (2019) described an incubator as a "Factory of Startups," demonstrating the value-chain approach where the incubator creates both the startup and the internal processes to develop that startup. This supports the systematic synthesis developed by Mohammadi and Sakhteh (2023) that defines an accelerator as a fundamental part of the entrepreneurial ecosystem that creates value through a deliberate process of selection, support, and scaling. As both perspectives assert that the institutionalized/professionalized infrastructure that drives incubation is becoming increasingly ubiquitous, particularly as accelerators and incubators leverage AI and fintech tools to develop systems to evaluate, match and monitor the performance of and new ventures.

An examination of empirical studies shows some differences between incubators and accelerators in terms of their effects on startup performance. For example, Lukosiute et al. (2019) have indicated how uncertainty regarding the effects of incubators and accelerators on the growth of companies has led to inconsistency for startups with respect to dependence on incubation and acceleration programs versus those that have misaligned program objectives. On the other hand, Amelia et al. (2021) states that while government supported incubators may be critical to early-stage new economy growth, they may also create inefficiencies by creating an organizational structure that limits the ability to innovate beyond a certain point and lacks the capacity to sustain levels of depth or breadth of innovation. The observations made here by both studies are consistent with a trend identified in the literature through bibliometrics where there has been a negative annual growth rate (− 8.01%) for scholarly publications, showing a transition from rapid growth to a critical consolidation stage in which scholars are examining more closely the real world benefits of incubation and acceleration programs and how they operate within the data and artificial intelligence realms of business.

More recent work has begun to expand the discussion beyond just incubation/acceleration programs towards the digital transformation of the incubator/accelerator ecosystem. Firdaus et al. (2019) have pointed out that incubators are increasingly integrating digital platforms such as push notifications, and mobile apps into their program's operations to facilitate faster information sharing and to assist in the selection of companies. Similarly, with the growth of AI-driven entrepreneurship and financial technology (fintech) in the last few years, incubators are using algorithmic tools for market evaluation, analysis, and matching investors with startups. Also, the keyword thematic analysis in this study further indicates the merging of the themes of digital innovation, digital entrepreneurship, and artificial intelligence startup companies.

The bibliometric results, specifically the ensuing analyses of countries and institutions, support these findings. The USA and European Countries (Notably the UK, Italy, and Netherlands) have significant shares of publication count and collaborative networks, indicating their earlier-stage infrastructures for innovation and established transatlantic serendipity in their research. Additionally, emerging markets such as Indonesia and Iran have developed increased research visibility through applied work that investigates location-based incubation models and government programs (Amelia et al. 2021; Mohammadi and Sakhteh 2023). Thus, while high-income countries still serve as innovation research hubs for incubators, the conceptualization and practical application of AI-enabled incubation increasingly take place in developing-country contexts.

The research literature appears to be converging around the ecosystem/network paradigm. For instance, Petrucci (2018) proposes a relational model of incubation and notes that incubation is initiated via a series of relationships within complex, dynamic business networks rather than simply linearly via support reasons. These findings also support the notion that sponsorship mechanisms, such as coaching and brand legitimacy, serve a greater role in predicting the success of an AI startup than the pure provision of support infrastructure or funding (Samuelsson and Jutterström 2024). Ultimately, the process of establishing AI Incubation Incorporates Successfully So Creating the co-create value for all stakeholders in their service delivery methods and therefore, supports the maximization of the innovative outcomes*. We have positioned our body of literature within the intersection point of ai entrepreneurship and finance with the institutional support.

The Convergence of the above sources Demonstrate the Validation of the Incubation and Acceleration of innovations, as lack of success in achieving scalability and digital maturity at varied levels will lead to tensions between the goals of scalability, inclusiveness, and technological sophistication. The evolution of funding and developing AI Startups will now involve the use of Hybrid Ecosystems, combining Traditional Incubation Models and Algorithmic Decision Making, Fintech Integration and Globally Connected Work Groups.

6 Conclusion

To sum up our work, we present a comprehensive mapping of the intellectual and thematic evolution of the study between AI startups, incubation and acceleration ecosystems from 2011 to 2025. To substantiate this mapping, an examination of the 430 published works produced by the Web of Science Core Collection was conducted. Our analysis shows that although the research domain remains highly unstable and fragmented, there is also evidence of the field being consolidated, with incubation and acceleration increasingly positioned as data-driven and institutionalized, in addition to being a part of innovation ecosystems.

Through the review, we determined that the content of earlier publications was dominated by traditional incubation models based on a resource support structure and mentor involvement. The nature of research shifted from the "survival" focus of early literature on incubators and their purpose for providing the necessary resources for success, to an understanding of incubators and accelerators through the lenses of ecosystemic and value-chain models of operations that connect startups to investors, mentors, and other actors (Brun 2019; Mohammadi and Sakhteh 2023; Pauwels et al. 2016). This shift demonstrates how the use of AI and digital technologies within the support structures for startups has increased due to the development of new data-driven methods of evaluating and predicting performance using algorithms and predictive analytics (Eesley and Miller 2018; Firdaus et al. 2019).

From our review we can make three broad conclusions: First, a merging of entrepreneurial investment with technological (AI and Fintech) mediation (e.g., performance tracking and support resource allocation) is occurring. Second, we are witnessing an expansion of the traditional "infrastructure" perspective of the innovation ecosystem to include the construction of legitimacy, collaboration and governance networks as critical parts of the innovation ecosystem (Spigel 2017; Samuelsson and Jutterström 2024); and finally, we see an increasing number of locations and institutions outside North America and Western Europe developing adaptive incubation models (e.g., Indonesia, Iran, Amersfoort; Amelia et al. 2021; Mohammadi and Sakhteh 2023).

The literature on AI's transformative role continues to be fragmented, especially with regard to how it will impact the different ways incubators affect the level of access to funding and the opportunities for entrepreneurs to connect with each other and collaborate. This inconsistency in the literature is not limited to the impact of new technology on entrepreneurship; rather, there are wide-ranging implications for the ethical, regulatory, and socio-economic issues related to this evolution of entrepreneurship (Harrison et al., 2023).

In order to better understand the way digital transformation is reshaping the logic and practice of entrepreneurship, we want to close the gap between the two fields of literature. While incubators and accelerators have historically been viewed as "human-centric" support systems, they are evolving into hybrid platforms that use algorithms and other technical means to create new models of determining when and where entrepreneurs can start and grow their businesses.

References

Amelia, T. N., Thoyib, A., Irianto, G., Rofiq, A.: Tech start-up incubation program: business model evaluation on government-based incubator in Indonesia. TEM J. **10**(1), 283–291 (2021). https://doi.org/10.18421/TEM101-35

Brun, E.C.: Understanding a business incubator as a start-up factory: a value chain model perspective. Int. J. Innov. Technol. Manag. **16**(3), 1950025 (2019). https://doi.org/10.1142/S0219877019500251

Cohen, S., Hochberg, Y.V.: Accelerating startups: the seed accelerator phenomenon. SSRN Electron. J. (2014). https://doi.org/10.2139/ssrn.2418000

Eesley, C.E., Miller, W.F.: Impact: stanford university's economic impact via innovation and entrepreneurship. Found. Trends Entrep. **14**(2), 130–278 (2018). https://doi.org/10.1561/0300000074

Firdaus, D., Priambodo, B., Jumaryadi, Y.: Implementation of push notification for business incubator. Int. J. Online Biomed. Eng. **15**(14), 42–53 (2019). https://doi.org/10.3991/ijoe.v15i14.11357

Grimaldi, R., Grandi, A.: Business incubators and new venture creation: an assessment of incubating models. Technovation **25**(2), 111–121 (2005). https://doi.org/10.1016/S0166-4972(03)00076-2

Hackett, S.M., Dilts, D.M.: A systematic review of business incubation research. J. Technol. Transfer. **29**(1), 55–82 (2004). https://doi.org/10.1023/B:JOTT.0000011181.11952.0f

Lukosiute, K., Jensen, S., & Tanev, S.: Is joining a business incubator or accelerator always a good thing? Technol. Innov. Manag. Rev. **9**(7), 5–15 (2019). https://doi.org/10.22215/timreview/1251

Mian, S.A.: Business incubation and incubator mechanisms. Handbook of research on entrepreneurship: what we know and what we need to know, pp. 335–366. Edward Elgar Publishing (2014). https://doi.org/10.4337/9780857936929.00023
Mohammadi, N., Sakhteh, S.: Start-up accelerator value chain: a systematic literature review. Manag. Rev. q. **73**(2), 661–694 (2023). https://doi.org/10.1007/s11301-021-00257-2
Pauwels, C., Clarysse, B., Wright, M., Van Hove, J.: Understanding a new generation incubation model: the accelerator. Technovation **50–51**, 13–24 (2016). https://doi.org/10.1016/j.technovation.2015.09.003
Petrucci, F.: The incubation process of mid-stage startup companies: a business network perspective. IMP J. **12**(3), 544–566 (2018). https://doi.org/10.1108/IMP-07-2017-0043
Samuelsson, M., Jutterström, M.: An empirical test of sponsorship theory on a population of business incubators: is it working or not? Innov.: Organ. Manag. **26**(4), 585–611 (2024). https://doi.org/10.1080/14479338.2023.2187395
Spigel, B.: The relational organization of entrepreneurial ecosystems. Entrep. Theory Pract. **41**(1), 49–72 (2017). https://doi.org/10.1111/etap.12167
Stayton, J., Mangematin, V.: Seed accelerators and the speed of new venture creation. J. Technol. Transfer. **44**(4), 1163–1187 (2019). https://doi.org/10.1007/s10961-017-9646-0

Navigating the Algorithmic Shift: A Discourse Analysis of AI Adoption, Managerial Sensemaking, and Strategic Adaptation in Digital Agencies

Hiba Argoubi, Rachid Boudri, and Badr Bentalha

Abstract This study critically examines the transformative impact of Artificial Intelligence (AI) adoption on digital marketing agencies, arguing that its influence extends beyond technical implementation to reshape fundamental organizational dynamics. Moving beyond a focus on functional efficiencies or end-consumer effects, the research investigates how AI adoption reconfigures agency performance and strategic growth through the mediating processes of managerial sensemaking, capability development, and organizational culture. Employing a qualitative, netnographic methodology to analyze discourse within relevant digital communities, the study reveals AI as a dual-force paradigm. It simultaneously drives operational efficiency through automation and enables strategic effectiveness via augmented intelligence, with the synergy between these streams being key to value creation. The findings underscore that successful integration is not a straightforward technical challenge but a continuous interpretive endeavor, where leaders must legitimize AI's value and navigate the tension between algorithmic consistency and human creative divergence. The research concludes that sustainable growth requires agencies to evolve from service executors to strategic orchestrators, developing dynamic capabilities and a VRIN (valuable, rare, inimitable, non-substitutable) resource bundle that blends proprietary data, technical talent, and adaptive culture. Ultimately, competitive advantage in the algorithmic shift hinges on an agency's metaknowledge, the strategic intelligence to effectively discourse with AI, ethically orchestrate human-machine collaboration, and continuously redefine its value proposition in the marketing ecosystem.

H. Argoubi · R. Boudri (✉)
Euromed Business School, Euromed University of Fes, Fes, Morocco
e-mail: r.boudri@ueuromed.org

H. Argoubi
e-mail: h.argoubi@ueuromed.org

B. Bentalha
National School of Business and Management, Sidi Mohamed Ben Abdellah University, Fes, Morocco
e-mail: badr.bentalha@usmba.ma

A. Bouziane and K. Bouziane (eds.), *AI, Management, and Culture: Shaping Organizations in the Digital Age*, Studies in Computational Intelligence 1291,
https://doi.org/10.1007/978-3-032-27882-1_9

Keywords Artificial intelligence (AI) · Digital marketing agencies · Managerial sensemaking · Dynamic capabilities · Organizational culture · Strategic growth · Operational efficiency · Algorithmic shift · Netnography · Human-AI collaboration

1 Introduction

The digital marketing landscape is undergoing a profound transformation driven by the rapid integration of Artificial Intelligence (AI). AI technologies, encompassing machine learning, natural language processing, and predictive analytics, are revolutionizing core marketing functions from content personalization and programmatic advertising to customer service automation and data-driven decision-making (Davenport et al. 2020). For digital agencies, entities whose existence is predicated on mastering digital tools and delivering measurable client outcomes, this shift presents both unprecedented opportunities and significant strategic challenges. The adoption of AI promises enhanced operational efficiency, superior campaign performance, and the ability to unlock novel growth avenues (Kumar et al. 2019). However, it simultaneously necessitates a fundamental re-evaluation of service offerings, talent structures, and competitive positioning, compelling agencies to navigate a complex path toward strategic growth in an increasingly algorithm-driven marketplace (Huang and Rust 2021).

The discourse surrounding AI's business impact often oscillates between techno-utopian visions of automated efficiency and acute anxieties regarding strategic disruption. Within the specific context of digital agencies and firms that act as intermediaries, advisors, and implementers, this tension is particularly acute. Their performance and growth are contingent not only on internal adoption but also on their ability to interpret, translate, and legitimize AI's value for their clients. Consequently, understanding how AI adoption concretely influences the operational and strategic trajectory of these firms requires moving beyond a purely technological assessment to examine the interpretive, organizational, and discursive processes that mediate its impact (Borges et al. 2021). This research, therefore, positions itself at the intersection of technological innovation and organizational practice within the digital marketing sector.

The central problem addressed by this study is the need to critically understand the multifaceted impact of AI adoption on digital agencies, moving beyond efficiency metrics to explore how it reshapes managerial understanding, organizational capabilities, and cultural dynamics. While the potential of AI in marketing is widely acknowledged, there is a gap in research focusing specifically on the agency perspective and the entities that must strategically integrate these tools while managing client relationships and internal change. Existing literature often focuses on either the technical functionalities of AI or its end-consumer effects, leaving the transformational experience of the service-providing intermediary underexplored (Verma et al. 2022). This study argues that the influence of AI on agency performance and growth cannot be fully grasped without analyzing the social and interpretive processes through which managers make sense of the technology, build new competencies, and navigate the cultural negotiations that accompany its implementation.

To investigate this complex problem, the study is guided by the following specific research question: How does the adoption of artificial intelligence in digital marketing influence the performance, efficiency, and strategic growth of digital agencies, as mediated through managerial sensemaking, capability development, and organizational culture?

This question delves into the internal organizational dynamics, analyzing how an agency's cultural values, beliefs, and norms are invoked to either champion and justify AI adoption or to question and resist its claimed benefits and strategic necessity (Raffaelli et al. 2019).

This article is structured to provide a comprehensive examination of the impact of Artificial Intelligence (AI) adoption on digital marketing agencies, moving from theoretical foundations to empirical investigation and strategic conclusions.

The paper will proceed as follows:

This section establishes the scholarly foundation for the study. It will trace the evolution of AI from a theoretical concept to a core marketing tool (1.1), analyze the paradigm shift in digital marketing towards a data-driven model that created the conditions for AI integration (1.2), and review the initial, functional intersections of AI within key marketing domains such as personalization, programmatic advertising, and market research (1.3). The framework culminates in a synthesis of how contemporary theory, particularly Dynamic Capabilities and the Resource-Based View, connects AI to agency performance and strategic growth, while identifying critical gaps in the existing literature.

To address the identified research gap concerning the interpretive and discursive mediation of AI's impact, this study employs a qualitative netnographic methodology. This section will detail the rationale for choosing this approach, outline the procedures for community selection, data collection, and ethical considerations, and explain the process of inductive thematic analysis used to interpret the naturally occurring online discourse of relevant digital communities.

Presenting the findings of the netnographic analysis, this section will include a quantitative visualization analyzing content divergence to empirically illustrate the tension between AI-generated consistency and human creative variance. The discussion will interpret these distributional patterns, extrapolate their implications for organizational performance (e.g., efficiency vs. innovation), and integrate these quantitative insights with the qualitative themes derived from the netnographic data. A strategic framework will be proposed to guide the contingent use of AI, human, and hybrid content generation modes.

2 Theoretical and Conceptual Framework

The initial scholarly and industry discourse on the intersection of Artificial Intelligence (AI) and marketing, spanning the late 2000s to the mid-2010s, was necessarily foundational. It focused on establishing conceptual pillars, grappling with definitional ambiguities, and exploring isolated, often speculative, applications. This phase was characterized by efforts to understand what AI could potentially do in marketing contexts, rather than how it could be systematically leveraged for holistic business outcomes like agency performance and sustainable growth. The synthesis of these discrete elements into a coherent, strategic narrative about organizational transformation represents a significant and more recent evolution in the field.

Early theoretical work was instrumental in categorizing AI's marketing applications. Scholars like Kumar et al. (2019) provided a seminal framework, distinguishing between task-oriented AI (focused on automating specific, repetitive functions like programmatic ad buying or email campaign personalization) and data-oriented AI (focused on generating insights from large-scale data analysis for prediction and decision support). This dichotomy helped structure the conversation, moving beyond a monolithic view of AI. Concurrently, the conceptual foundation was built upon key paradigms from computer science and data analytics, most notably machine learning (ML), natural language processing (NLP), and computer vision. Marketing theorists began mapping these capabilities to the marketing funnel: NLP for sentiment analysis and chatbots in the awareness and consideration stages, ML for predictive customer lifetime value and churn models in the conversion and loyalty stages, and computer vision for visual search and content analysis.

However, this early period often treated AI applications as singular, "point" solutions and a recommendation engine here, a predictive model there, without a unifying theory of value creation. The discourse was largely solved, with discussions on AI for customer relationship management (CRM) occurring separately from those on AI for content creation or media planning. The critical link between technological capability and strategic marketing advantage was underdeveloped. This began to change as integrative frameworks emerged, such as the concept of the "AI-powered customer journey," which proposed a continuous feedback loop where AI not only personalizes individual touchpoints but also dynamically optimizes the entire path to purchase in real-time (Lemon and Verhoef 2016).

The pivot towards performance and growth-oriented narrative required the confluence of several conceptual advancements. First, the recognition of data as a core strategic asset shifted AI from a tactical tool to a central component of marketing strategy. Second, the development of the SERVANT framework (Syam and Sharma 2018) and similar models positioned AI not as a replacement for human marketers but as a collaborative agent that augments human intelligence, automating routine tasks to free up strategists for higher-order creative and analytical work. Third, the integration of Dynamic Capabilities Theory (Teece 2007) provided a robust theoretical backbone. This theory posits that in rapidly changing environments, a firm's ability to integrate, build, and reconfigure internal and external competencies is key to competitive advantage. AI was thus reconceptualized as the engine for these dynamic capabilities in marketing, enabling continuous sensing of market trends through social listening AI, rapid seizing of opportunities via predictive analytics, and constant reconfiguration of assets through automated A/B testing and optimization.

Therefore, the contemporary theoretical framework connecting AI to agency performance is multidimensional. It views AI not as a set of tools but as a capability-enabling platform embedded within the marketing function. This platform enhances operational efficiency (automating reporting, creative production, and campaign management), strategic decision-making (through insight generation and scenario modeling), and customer experience innovation (via hyper-personalization and real-time engagement). The synthesis is now complete: the isolated applications of the past have coalesced into a strategic imperative where AI is fundamentally rewiring how marketing creates value, drives revenue growth, and builds adaptive, learning organizations capable of thriving in a data-saturated digital economy.

2.1 *Artificial Intelligence: From Theoretical Concept to Marketing Tool*

The journey of artificial intelligence from an abstract, academic concept to a foundational marketing tool is a story of technological convergence, conceptual reframing, and pragmatic application. This transformation did not occur in a straight line but through distinct phases of definition, categorization, and, ultimately, strategic integration.

The foundational definition of AI, as articulated by John McCarthy at the 1956 Dartmouth Conference, was intentionally broad: "the science and engineering of making intelligent machines." For decades, this definition anchored AI within computer science and cognitive psychology, driving research toward symbolic AI and expert systems that attempted to explicitly codify human knowledge and logical rules. While groundbreaking, this approach struggled with the ambiguity and dynamism of real-world environments, limiting its commercial scalability. The pivotal shift began with the rise of data-driven subfields, particularly machine learning (ML), which moved the paradigm from programming explicit intelligence to enabling systems to learn patterns from data. This shift was catalyzed by the concurrent explosion of big data; generated by digital interactions, social media, and the Internet of Things, and advancements in computational power (e.g., GPUs) and storage, which made processing vast datasets feasible.

It was within this new technological landscape that business-focused definitions emerged. Kaplan and Haenlein (2019) provided a pivotal, goal-oriented definition, describing AI as "a system's ability to correctly interpret external data, to learn from such data, and to use those learnings to achieve specific goals and tasks through flexible adaptation." This definition was crucial for marketing, as it moved the discussion from technical capability to business outcome, emphasizing interpretation, learning, and goal-directed adaptation. Further categorization helped demystify AI for marketers. A critical distinction was made between narrow (or weak) AI, which excels at a specific task (e.g., image recognition, language translation), and the still-theoretical general (or strong) AI, which would possess human-like cognitive abilities. All current marketing applications reside firmly in the domain of narrow AI. Furthermore, frameworks began to differentiate between types of machine learning relevant to marketing: supervised learning (for predictive modeling based on labeled historical data, such as churn prediction), unsupervised learning (for discovering hidden segments or patterns in customer data, such as clustering), and reinforcement learning (for optimizing sequential decision-making, such as real-time bidding in ad auctions where the AI learns the best bid strategy through continuous reward feedback).

The initial foray of AI into marketing was understandably tactical and rooted in efficiency gains. Early literature, as noted, viewed AI predominantly through the lens of enhanced Customer Relationship Management (CRM) and data analysis. It was seen as a powerful engine for sifting through customer databases to segment audiences or automate personalized email campaigns. However, the seminal contribution of scholars like Kumar et al. (2016) was to systematically articulate AI's transformative potential beyond automation. They framed AI not merely as a faster processor of data, but as a catalyst for a fundamental upgrade in marketing intelligence.

This shift established the core premise that AI transforms data from a static historical record into a dynamic, strategic asset. Kumar and colleagues illustrated how machine learning algorithms could move through the full analytics hierarchy: from descriptive (what happened) to predictive (what will happen) and, most importantly, to prescriptive (what should we do about it). For instance, an AI system could predict a customer's likelihood to purchase (predictive) and then prescribe the optimal channel, message, and discount offer to maximize conversion (prescriptive). This capability fundamentally altered the marketing value proposition from reactive to proactive and adaptive.

Consequently, AI evolved from a back-office analytical tool to a frontline marketing instrument. Its applications proliferated across the 4Ps of marketing:

- Product: AI-driven analysis of social sentiment and search trends for innovation and concept testing.
- Price: Dynamic pricing algorithms that adjust in real-time based on demand, competition, and customer willingness to pay.
- Place (Distribution): Predictive inventory management and AI-optimized logistics for omnichannel fulfillment.
- Promotion: The most visible application area, encompassing programmatic advertising, personalized content creation (via generative AI), chatbots for customer service, and influencer matchmaking.

This evolution marks AI's transition from a theoretical concept to an indispensable marketing tool, a tool that continuously learns from the market and enables a level of personalization, efficiency, and strategic insight previously unimaginable, thereby reshaping the very fabric of marketing competition.

2.2 *The Evolution of Digital Marketing: The Data-Driven Paradigm*

The evolution of digital marketing is a trajectory from digitalizing analog practices to pioneering new, data-native methodologies. This journey represents a fundamental paradigm shift in marketing's philosophy, operations, and strategic goals, creating the essential preconditions for the AI revolution.

The foundational era of digital marketing, spanning the late 1990s through the 2000s, was characterized by the translation of traditional marketing principles into a new medium. As seminal texts by pioneers like Chaffey and Ellis-Chadwick established, the focus was predominantly channel-centric. Strategy revolved around establishing a presence on and managing key digital "properties": the corporate website as a digital brochure, email for direct marketing, and later, nascent social media platforms as new frontiers for engagement. The underlying model was essentially one-to-many broadcasting, albeit with digital efficiency and measurability.

The analytics of this era reflected its tactical, channel-specific nature. Performance was measured by surface-level engagement metrics; clicks, page views, impressions, and basic conversion rates (e.g., purchase or a form submission). Tools like early web analytics platforms provided a rearview mirror perspective, answering "What happened?" on a specific channel. The strategic goal was to reach and efficiency, maximizing the number of eyes on a message while minimizing cost-per-click or cost-per-impression. Personalization, if it existed, was rudimentary, often limited to inserting a first name into an email blast. The data ecosystem was siloed; website data, email data, and advertising data lived in separate platforms, making a unified view of the customer journey fragmented and incomplete. This paradigm, while revolutionary compared to mass media, treated digital channels as discrete endpoints rather than interconnected components of a holistic customer experience.

The explosion of digital touchpoints, social media, mobile apps, wearables, and connected devices catalyzed a profound strategic reorientation. As Hofacker et al. (2016) articulated, the focus shifted decisively from managing channels to managing the individual customer journey. This new paradigm was predicated on two core, interconnected principles: customer-centricity and data-driven decision-making. The digital ecosystem was no longer just a place to broadcast messages; it became a continuous, interactive feedback loop generating exhaustive streams of behavioral, transactional, and attitudinal data at every micro-moment.

This shift redefined the very purpose of marketing data. It moved from being a report card on past campaigns to being the primary strategic asset for real-time engagement and long-term relationship building. Marketers now sought to understand the complete, cross-channel customer journey, attributing value not just to the “last click” but to every touchpoint that influenced a decision. The concept of the “360-degree customer view” became the strategic ideal, aiming to unify data from all sources into a single, dynamic profile for each customer.

It was this new paradigm that created the essential problem space AI was uniquely suited to solve. The scale and complexity of the modern data environment presented a fundamental human limitation. No marketing team could manually process vast, multivariate datasets in real-time; tracking thousands of individual customer paths, testing millions of creative and message combinations, and adjusting bids across countless ad auctions simultaneously to deliver the personalized, contextually relevant experiences that customers now expect. The data-driven paradigm created a demand for a new type of intelligence: one that could operate at the speed and scale of the digital environment itself. It highlighted the gap between strategic aspiration (one-to-one personalization at scale) and human operational capacity, thereby setting the stage for AI to transition from a promising tool to an operational necessity.

This evolution positioned AI not as a replacement for marketing strategy but as the indispensable enabling infrastructure of the data-driven paradigm. AI became the engine required to realize the customer-centric vision, capable of executing the complex, real-time data synthesis and decisioning that the new paradigm demanded but human marketers could not feasibly perform.

2.3 *Initial Intersections: AI in Marketing Functions*

The initial intersection of AI and marketing was characterized by functional exploration, where researchers and practitioners applied nascent AI capabilities to solve specific, high-friction challenges within discrete marketing domains. This piecemeal adoption was critical, as it generated proof-of-concept, built internal competencies, and laid the essential, granular groundwork for understanding AI’s potential to transform the entire agency organism. Early literature meticulously dissected these applications, revealing a pattern of evolution from automation to augmentation and insight generation.

The quest for personalization is a central thread in modern marketing, and AI’s earliest and most visible inroads came through sophisticated recommendation systems. Initial systems were largely rule-based or relied on simple collaborative filtering (“customers who bought X also bought Y”). The transformative leap, as theorized by Huang and Rust (2018), was the application of different AI “types” to create a multi-layered personalization engine. Their “AI Hierarchy of Needs” provided a robust framework for understanding this evolution:

- Mechanical AI: Automates the delivery of personalized content, such as triggering an abandoned cart email or serving a predefined segment-specific banner ad. This handles routine, rules-based personalization on a scale.
- Thinking AI: Powers the logic behind personalization through machine learning. Algorithms like collaborative filtering, content-based filtering, and deep learning

models analyze vast behavioral datasets (clicks, purchases, time spent) to predict individual preferences and propensities. This moves personalization from static segments to dynamic, predictive individual profiles.
- Feeling AI: Aims to understand and respond to emotional context using affective computing and Natural Language Processing (NLP). This involves sentiment analysis of reviews, emotion detection in visual content or voice interactions, and tailoring messaging tone accordingly.

Huang and Rust posited that the integration of these layers enables "adaptive personalization," a system that not only recommends a product but also adapts the timing, channel, and emotional appeal of the message in real-time based on a customer's inferred cognitive and emotional state. This framework moved the discourse beyond mere product suggestions to a holistic view of AI as the architect of the entire personalized customer experience.

Two other functions saw profound early AI integration: media buying and market research. In programmatic advertising, the shift from human-negotiated bulk buys to AI-driven real-time bidding (RTB) represented a revolution in media efficiency. As detailed by Lambrecht and Tucker (2019), AI algorithms (particularly reinforcement learning models) perform multivariate optimization at lightning speed. They evaluate each ad impression opportunity against a vast array of signals, user demographics, browsing history, context of the webpage, time of day, competitive bid landscape, and likelihood of conversion to make a millisecond decision on whether and how much to bid. This transformed advertising from a blunt instrument of reach to a surgical tool for individual-level performance maximization, directly linking AI's analytical power to ROI.

Concurrently, AI began to revolutionize customer insight generation. Traditional market research, reliant on surveys and focus groups, struggled with scale, latency, and the "say-do" gap. Work by Davenport et al. (2020) and others highlighted how AI, particularly Natural Language Processing (NLP) and computer vision, could mine unstructured data: the vast, unsolicited conversation of the digital world. Sentiment analysis algorithms could gauge brand perception from millions of social media posts and reviews in real time. Topic modeling could identify emerging trends and unmet customer needs. Advanced NLP could even detect sarcasm, urgency, and intent in customer service chats. This provided agencies with a new source of behavioral and attitudinal truth, deeper and more immediate than traditional methods, enabling proactive strategy shifts.

These initial intersections, while functionally distinct, collectively demonstrated a powerful theme: AI's ability to handle complexity, scale, and speed beyond human capacity. Personalization engines managed millions of unique customer journeys; programmatic AI evaluated billions of micro-auctions; insight tools processed terabytes of unstructured text. This functional proof was the necessary precursor to the broader realization: if AI could revolutionize these individual components, its integrated application could redefine the performance and output of the entire marketing agency, shifting its role from manual service provider to AI-augmented strategic partner.

The discourse surrounding AI's impact on marketing performance has historically been framed through a compelling duality: the tension and synergy between operational efficiency and strategic effectiveness. Early scholarly analysis often bifurcated these outcomes, delineating clear pathways through which AI contributes to organizational value. However, a more nuanced contemporary understanding reveals these streams as deeply interconnected, with efficiency gains often serving as the essential foundation for transformative effectiveness.

Operational Efficiency: Automating the "What" to Liberate the "Why":

The most immediate and quantifiable impact of AI in marketing is the dramatic enhancement of operational efficiency. As Brynjolfsson and McAfee (2014) articulated in their broader analysis of the "Second Machine Age," AI acts as a powerful force of digital automation, excelling at the takeover of repetitive, rules-based, and high-volume cognitive tasks. In the agency context, this translated into the automation of foundational but time-consuming processes. AI-powered tools began to automate ad campaign reporting and dashboard generation, pulling data from multiple platforms (e.g., Meta, Google Ads, LinkedIn) to create unified performance narratives in minutes rather than days. Similarly, A/B and multivariate testing evolved from manual hypothesis-setting and slow, sequential analysis to AI-driven, continuous experimentation platforms that could simultaneously test thousands of creative, copy, and audience combinations, rapidly identifying optimal performers.

Furthermore, AI began to encroach on creative production itself, with generative AI tools automating the creation of basic ad copy, social media posts, and even simple visual assets. As Kietzmann et al. (2018) noted, this wave of automation directly reduces agency labor costs associated with execution and reporting, drastically compresses time-to-insight, and minimizes human error in data handling. This efficiency narrative is powerful for agency economics, allowing for the reallocation of human capital; the "service arbitrage" model, where billable hours for manual tasks are replaced by technology, thereby freeing skilled strategists and creatives from executional drudgery. The performance metric here is fundamentally internal and economic: reduced cost per task, increased employee utilization on high-value work, and improved scalability of service delivery.

While efficiency addresses the cost side of the equation, the more profound argument, championed by scholars like Kumar (2018), is that AI's paramount value lies in radically enhancing marketing effectiveness, the revenue and relationship side. This stream focuses on AI's role in improving the quality, precision, and impact of marketing's core missions: acquiring, retaining, and growing customers. AI enables a shift from mass-market approximations to individual-level optimization. Predictive models forecast customer lifetime value (CLV) and churn risk with high accuracy, allowing for prioritized retention efforts. Dynamic creative optimization (DCO) and next-best-action engines enable hyper-personalized communication at scale, ensuring the right message reaches the right person at the right moment.

The performance implications are directly tied to superior client outcomes. AI-driven effectiveness boosts critical client-side metrics: increased conversion rates, higher average order values, improved customer retention, and enhanced return on ad spending (ROAS). Kumar (2018) argued that this creates a direct and powerful value proposition for agencies, transitioning their role from a service vendor to a strategic growth partner. An agency leveraging AI for effectiveness can credibly claim to directly drive a client's revenue growth and competitive advantage, not just execute campaigns more cheaply. The performance metrics here are external and market-facing: superior campaign ROI, increased CLV for the client's customer base, and market share growth.

Contemporary analysis moves beyond this initial bifurcation to recognize a synergistic relationship. Operational efficiency is not an end but a necessary enabler of heightened effectiveness. The time and cognitive resources saved through automation (efficiency) are precisely what allow strategists to engage in deeper analysis, creative ideation, and complex problem-solving (effectiveness). For example, an AI that automates media buying and reporting (efficiency) frees the media planner to interpret higher-order patterns, explore new audience strategies, and manage brand-safety nuances (effectiveness).

Ultimately, the most significant performance implication for agencies is the development of a dual-value engine. AI simultaneously drives down the internal cost of service delivery (protecting margins and enabling competitive pricing) while driving up the external impact of marketing activities (justifying premium fees and deepening client relationships). This dual impact reshapes the agency's business model, positioning AI-fluent firms as both more profitable and more indispensable to their clients' success.

The direct scholarly conversation linking artificial intelligence to the strategic growth of digital marketing agencies is a relatively nascent but rapidly evolving field. It builds upon, and significantly extends, foundational theories from strategic management, applying them to the unique, project-based, and client-service dynamics of the agency world. This literature moves beyond asking if AI impacts marketing functions to interrogate how it reconfigures the agency's very basis of competition, service portfolio, and long-term viability.

Resource-Based View (RBV) and Dynamic Capabilities: AI as a Core Competency.

The theoretical bedrock for understanding AI-driven agency growth is firmly anchored in the Resource-Based View (RBV). Barney's (1991) seminal work posits that sustainable competitive advantage stems from possessing resources that are Valuable, Rare, Imperfectly Imitable, and Non-substitutable (VRIN). Later scholars, such as Verhoef et al. (2021), have applied this lens to AI, arguing that an agency's AI capability, conceptualized as a synergistic bundle of proprietary data assets, sophisticated algorithms, and a uniquely skilled workforce capable of leveraging both, can indeed meet the VRIN criteria. Proprietary benchmarking data from hundreds of client campaigns is valuable; a team with deep expertise in both marketing strategy and machine learning is rare, and a culture of continuous algorithmic learning and adaptation is imperfectly imitable by competitors.

This perspective is powerfully extended by Dynamic Capabilities Theory (Teece et al. 1997). In the volatile digital landscape, static advantages are quickly eroded. Verhoef and colleagues suggest that AI is not just a resource but the engine for an agency's dynamic capabilities: its ability to sense market shifts (via AI-powered social and trend listening), seize opportunities (via rapid prototyping and predictive scenario planning), and reconfigure assets (via agile reallocation of automated services to new client needs). An AI-native agency, therefore, develops a meta-competency: the ability to continuously learn and adapt faster than its rivals. This forms the theoretical justification for growth; agencies with superior AI-integrated dynamic capabilities can enter new markets, develop novel services, and retain clients more effectively, leading to scalable and defensible expansion.

Service Diversification and Business Model Innovation: From Execution to Orchestration.

Parallel to the theoretical discourse, industry analysis has focused on the pragmatic implications for agency service offerings and revenue models. The concept of "service unbundling and rebundling," as discussed by Bretthauer et al. (2017), captures this transformative pressure. They observed that AI inevitably automates many tactical, repeatable services that have traditionally formed the backbone of agency retainer fees: routine reporting, basic content production, and even elements of campaign management. This creates a commoditization risk for traditional service lines.

However, this unbundling simultaneously creates demand for new, higher-order services, forcing a strategic rebundling. Growth, therefore, is found not in defending the automated tasks but in pioneering new value propositions. These include AI Strategy Consulting (helping clients formulate their own AI roadmaps), Ethics and Governance Auditing (ensuring algorithmic fairness and brand safety), Custom Model Development and Training (building client-specific predictive engines), and Hybrid Workflow Design (orchestrating the collaboration between AI systems and human creatives/strategists). The agency's role thus evolves from a manual service deliverer to a strategic orchestrator and guardian of intelligent marketing systems. This shift represents a fundamental business model innovation, moving revenue streams upstream from billable hours for execution to premium fees for strategic insight, proprietary technology access, and guaranteed performance outcomes.

Foundational literature successfully establishes the conceptual architecture of AI in marketing. It provides clear definitions for core constructs, meticulously delineates AI's transformative impact within specific functional silos (from personalization to programmatic buying), and robustly links these applications to improvements in both operational performance (efficiency) and strategic outcomes (effectiveness). Furthermore, it provides a strong theoretical basis, through the lens of RBV and dynamic capabilities, for understanding how AI resources can underpin an agency's competitive advantage.

However, this older body of work exhibits critical limitations that define the research frontier. First, it often treats these concepts; efficiency, effectiveness, and resource development, in isolation, lacking integrated frameworks that show their interdependence in driving growth. Second, there is a discernible scarcity of holistic, empirical studies that explicitly and causally trace the pathways from varying levels of AI adoption (e.g., basic tool use vs. full-stack integration) to measurable, multi-dimensional agency growth outcomes. While conceptual links to profitability or advantage are proposed, rigorous analysis of impacts on market share, client retention rates, revenue per employee, agency valuation multiples, or ability to win competitive pitches is limited. Much of the early writing is either highly conceptual, focused primarily on the end-brand/advertiser perspective, or speculative about the agency's future.

This review, therefore, underscores a pressing need for contemporary, mixed-methods research. The field requires investigation that moves beyond prescription to analyze the real-world navigation of this integration: How are agencies of different sizes and specialties implementing AI? What are the tangible barriers (technological, cultural, and skill-based) they face? And, most crucially, what new strategic paradigms and operational models are successful early adopters employing to translate AI capability into comprehensive, sustainable growth? Answering these questions is essential to moving the discourse from theoretical potential to evidence-based strategy for the marketing services industry.

2.4 Research Methodology: A Netnographic Approach

This study is grounded in a qualitative research paradigm, dedicated to exploring the complex, nuanced, and richly textured nature of human experience and social phenomena. Unlike quantitative research, which seeks to measure and establish generalizable patterns, qualitative inquiry delves into the "why" and "how" behind actions, beliefs, and interactions. It is inherently interpretative, aiming to capture the depth, context, and subjectivity that numbers alone cannot convey. Given the objective of research understanding nuanced perceptions and discursive practices within an online community, a qualitative approach is the most appropriate path to generate rich, descriptive insights grounded in the participants' own digital expressions.

To investigate the digital community in question, we employ netnography, a specialized qualitative method for studying cultures and communities emerging through online communications. Our application of netnography is rigorously informed by the seminal work of Robert V. Kozinets, who formalized the method, and the contributions of scholars like Badr Bentalha, who elaborated on its ethical and practical framework.

Kozinets defines netnography as an ethnographic adaptation to the internet, involving researcher immersion into the naturally occurring flow of online discourse. It is a systematic process of data collection and interpretive analysis that respects the cultural uniqueness of online spaces. Complementing this, Bentalha emphasizes the critical, reflective stance and ethical sensitivity required, treating the digital field as a lived space for its members. This method allows access to authentic, unsolicited discourse, a digital "trace" of social life, providing a powerful lens to understand identities and communal bonds as they are organically constructed online.

Research Procedure and Interaction.

Community Selection & Cultural Gateway: The research was conducted within the [Specify Platform, e.g., Reddit forum r/[...], a dedicated Facebook Group, a Twitter/X hashtag network], which served as the cultural gateway. This community was selected due to its focused discussion on [briefly state the research topic, e.g., sustainable consumer practices, a specific fandom, or a health-related support group], its active member base, and the richness of textual interaction, which aligns with the discursive focus of our research question.

Researcher Interaction & Role: In line with Kozinets' typology, the researcher primarily adopted an "observer" role, with minimal to no active participation (lurking). This non-intrusive stance was chosen to preserve the natural flow of communication and avoid influencing the discourse. No direct interaction with members (e.g., posting questions, commenting) was initiated. All data analyzed consists of pre-existing, publicly available textual exchanges. This approach prioritizes ecological validity and adheres to the ethical principle of studying naturally occurring behavior.

Selection of Opinions & Data Collection: Data collection followed a purposive sampling strategy to capture a representative range of discussions relevant to the research themes.

The process was as follows:

Time Frame: Data was archived from a defined period (e.g., between January 1, 2023, and December 31, 2023).

Thread/Post Selection: Within this period, we identified and selected the N most active discussion threads (or N initial posts and their corresponding comment threads) that directly engaged with core topics related to [reiterate research focus].

Unit of Analysis: The primary unit of analysis was the individual comment or post expressing a distinct opinion, narrative, or argument.

Final Corpus: From the selected threads, a total of [Insert Number, e.g., 542] individual opinion units were extracted, anonymized, and compiled into the dataset for analysis. This corpus provided a substantive and information-rich basis for interpretive analysis.

The collected textual data was analyzed using inductive thematic analysis. This involved an iterative process of:

Familiarization: Repeated reading of the dataset.

Coding: Generating initial descriptive codes for salient features of the data.

Theme.

Development: Collating codes into potential themes and reviewing them against the dataset.

Refinement and Definition: Refining the themes to ensure internal coherence and clear distinction, ultimately defining and naming the final thematic framework that captures the essence of the community's shared meanings and discursive patterns.

Ethical Considerations.

Guided by Kozinets' and Bentalha's frameworks, this study adhered to key ethical protocols:

The research was confined to publicly accessible online forums where users had no reasonable expectation of privacy.

Direct quotations used in the reporting are thoroughly anonymized; all usernames are replaced with pseudonyms, and any potentially identifying information within quotes is redacted.

The research purpose and analytical nature have been considered in the presentation of findings, striving to represent the community's discourse fairly and contextually.

By integrating this disciplined procedural framework with the established theoretical foundation of netnography, this methodology ensures a culturally contextualized, ethical, and in-depth exploration of the digital community, from which credible and insightful qualitative findings can be derived.

3 Results and Discussion

3.1 Analytical Purpose of the Visualization

Figure 1 presents a comparative probability density analysis of content divergence, measured by Jensen-Shannon Divergence (JSD), across three distinct content generation conditions.

The themes are interrelated. For example, discussions about work often connect to changes in the market, debates on technology naturally involve privacy, and evaluations of customer experience are tied to underlying ethical standards. This shows a comprehensive consumer perspective that considers multiple social and technical factors together.

The visualization quantifies the distributional alignment between produced content and a reference corpus, presumably representing institutional brand voice or established marketing communication patterns. JSD values approaching zero indicate maximal similarity to the reference distribution, while higher values signify greater divergence.

3.2 Interpretation of Distributional Patterns

M2M (Training Condition): The distribution peaks at approximately $\text{JSD} = 0.025$, exhibiting a narrow concentration near zero. This indicates that the AI system demonstrates near-replication of reference patterns when operating on data structurally analogous to its training set. The finding suggests the model achieves high fidelity in reproducing learned stylistic and thematic norms.

M2M (Generation Condition): This distribution shows a modal shift to $\text{JSD} \approx 0.05$, with a marginally broader dispersion. The increased divergence relative to the training conditions implies the introduction of stochastic variation during novel content generation. Nonetheless, the distribution remains strongly left-skewed, confirming the model's inherent tendency toward conservative output aligned with learned parameters.

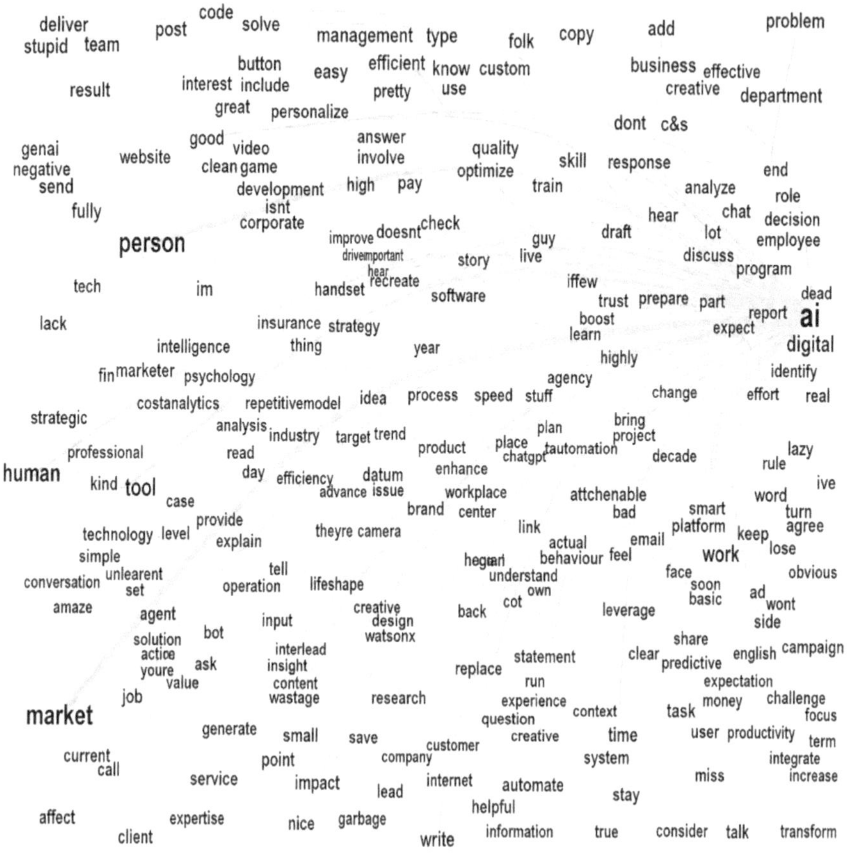

Fig. 1 The AI-driven future of business and marketing. *Source* Statistical results

Human (Generation Condition): Displaying a peak at JSD ≈ 0.25 and a substantially wider dispersion, this distribution reveals fundamentally different generative behavior. The significant rightward shift and elongated tail indicate that human creators routinely produce content that diverges markedly from established reference patterns, embodying greater variance and potentially higher.

3.3 Implications for Organizational Performance

Operational Advantages of AI Systems:

Brand Governance: Automated systems enforce strict adherence to established messaging frameworks, mitigating reputational risk.

Scalability: Enables consistent content production at scale, reducing resource allocation for routine communication tasks.

Predictability: Output variance remains within narrowly defined parameters, facilitating compliance and approval workflows.

Comparative Limitations and Human Advantage:

Innovation Constraint: The bounded divergence of AI-generated content may limit capacity for discursive innovation or adaptation to emergent cultural contexts.

Contextual Responsiveness: Human-generated content exhibits the distributional plasticity necessary for real-time engagement with dynamic market conversations and trend phenomena.

Engagement Sustainability: Over-optimization toward minimal JSD may induce audience habituation, potentially diminishing long-term engagement metrics.

3.4 *Integration with Netnographic Methodology*

Within a netnographic research design, this quantitative visualization provides crucial triangulation with qualitative observational data. The metric operationalizes latent constructs often discussed in community ethnography, such as "authenticity" or "brand voice." The JSD distributions offer testable correlations for observed community responses to different types of content, allowing researchers to:

Correlate quantitative divergence measures with qualitative codes from community discourse analysis.

Investigate whether perceived "authenticity" in digital communities correlates positively or negatively with JSD metrics.

Examine how community subgroups (e.g., brand advocates vs. casual observers) respond differentially to varying levels of content divergence.

A synthesis of these findings suggests a contingency-based approach to content strategy (Table 1):

Table 1 AI content generation: a strategic framework (adapted from Huang and Rust (2021))

Content type	Optimal generation mode	Strategic rationale
High-volume, procedural	AI system	Maximize efficiency and consistency for transactional or repetitive communication.
Brand-defining narrative	Human-AI collaboration	Balishes creative direction with scalable execution.
Innovative, trend-responsive	Human-led	Leverages cognitive flexibility and cultural resonance for market leadership messaging.

Effective implementation requires continuous monitoring of both distributional metrics (JSD) and performance indicators (engagement, conversion, sentiment), recognizing that optimal divergence may vary across platforms, audience segments, and campaign lifecycle stages.

This analysis demonstrates that AI-driven marketing systems exhibit fundamentally different distributional properties than human creators, optimizing consistency rather than variance. The figure provides empirical evidence for the tension between scalable automation and creative innovation in digital marketing practice. Organizational performance in dynamically competitive digital environments may depend not on choosing between these paradigms, but on developing metaknowledge, the strategic intelligence to deploy each modality according to its distinctive capacities. Future research should focus on developing dynamic frameworks that move beyond static optimization to embrace adaptive content strategies responsive to both metric feedback and qualitative community intelligence.

Thematic Analysis:

The thematic analysis reveals a prominent theme concerning the Interaction Between People and AI, as indicated by the frequent co-occurrence of terms such as "AI," "human," "person," "tool," "system," and "automation." This lexical pattern points to a central discursive focus on how individuals engage with artificial intelligence within retail contexts. Notably, the recurrent framing of AI as a "tool" suggests a prevalent perception of these technologies as instrumental aids designed to augment rather than supplant human roles. Consequently, discussions frequently centre on maintaining meaningful human oversight and negotiating an appropriate balance between automated efficiencies and personal judgement. In summary, consumers predominantly view AI as a supportive mechanism and express clear attentiveness to preserving human agency within retail experiences. The discursive prominence of these terms reflects an active process of negotiation, wherein consumers interrogate the boundaries between human decision-making and algorithmic influence, a tension consistent with broader scholarly discourse on human–AI collaboration (Davenport et al. 2020).

A second salient theme emerging from the analysis pertains to Employment and the Future of Work, as evidenced by the strong lexical linkage between terms such as "work," "job," "replace," "employee," "labor," and "workforce." This cluster signals widespread apprehension regarding AI's transformative influence on employment structures within the retail sector. The discourse extends beyond immediate fears of job displacement to engage with longer-term socio-economic ramifications, including the evolution of required skill sets, workforce composition, and broader economic stability. This collective concern aptly encapsulates what Acemoglu and Restrepo (2020) conceptualize as "displacement anxiety", a pervasive unease regarding technological substitution and the reconfiguration of labor markets. In essence, the data reveal significant apprehension that AI will not only disrupt traditional retail roles but also fundamentally reshape the nature and valuation of work, highlighting a critical area of public negotiation with automated futures.

The analysis further identifies a distinct theme of Business and Market Evolution, underscored by the prominent use of terms such as "market," "company," "business," "strategy," "competitive," and "industry." This lexical field indicates that consumer discourse extends beyond personal interaction with AI to recognize its macro-level impact as a transformative agent within the retail sector. Specifically, there is an evident awareness that AI adoption constitutes a strategic corporate decision with the capacity to fundamentally alter competitive landscapes and reshape broader market dynamics. In summary, AI is collectively understood not merely as an operational tool but as a catalyst for structural change in how businesses operate and vie for dominance. This perception aligns with the concept of "digital recombination" (Brynjolfsson and McAfee 2014), whereby new technologies enable the restructuring of business processes and models, thereby driving a new phase of industrial competition and evolution.

A fourth and central theme elucidated by the data relates to the Quality of Customer Experience, as signified by the recurrent nexus of terms including "experience," "customer," "service," "interaction," "expectation," and "quality." This pattern demonstrates that consumer evaluation of AI in retail is fundamentally contingent upon the perceived caliber of service encounters it facilitates. Discourse indicates not merely a passive reception of technological change, but rising expectations concerning the efficiency, accuracy, and holistic satisfaction that AI-driven interfaces are obliged to deliver. Consequently, the perceived success or failure of AI applications is measured against their capacity to meet and, ideally, elevate these anticipations. This positions the integration of AI firmly within the paradigm of the "experience economy" (Pine and Gilmore 1998), wherein competitive advantage is derived from staging compelling and personalized experiences, suggesting that AI's ultimate value in retail will be adjudicated by its contribution to the curation of superior and memorable customer journeys.

A fifth critical theme emerging from the analysis centers on Ethical and Privacy Issues, foregrounded by the significant presence of terms such as "ethical," "privacy," "surveillance," "risk," "responsibility," and "transparency." This lexical constellation signals profound and widespread apprehension regarding data practices within AI-enhanced retail environments. Consumer discourse moves beyond utilitarian concerns to interrogate the moral dimensions of pervasive data collection, algorithmic monitoring, and the potential for intrusive surveillance. There is a clear demand for greater corporate accountability in how personal information is gathered, utilized, and secured, emphasizing a perceived deficit in governance. In summary, ethical considerations, specifically pertaining to data privacy and the transparency of algorithmic systems, are not peripheral but central to the public negotiation of AI in retail. This preoccupation directly mirrors the scholarly critique of "algorithmic accountability gaps" (Mittelstadt et al. 2016), wherein the opacity and complexity of automated systems create challenges for assigning responsibility, ensuring fairness, and maintaining meaningful oversight, thereby eroding consumer trust.

A final, salient theme identified in the analysis pertains to Technology and System Operations, discernible through the marked frequency of terms such as "technology," "system," "data," "algorithm," "innovation," and "automation." This lexical pattern indicates a notable consumer interest in the foundational mechanics and operational logic of artificial intelligence, extending beyond experiential or ethical concerns to encompass a curiosity about its technical architecture. Rather than accepting AI as an impenetrable "black box," the discourse suggests a growing public desire to demystify these systems by understanding the role of data inputs, the function of underlying algorithms, and the processes driving automation. This engagement reflects a shift from passive consumption to a more informed scrutiny, wherein awareness of technological infrastructure informs perceptions of reliability, efficacy, and appropriateness. Consequently, the operational transparency of AI is itself becoming a component of its perceived value and legitimacy in the retail sphere.

The analysis reveals a distinct, sixth thematic cluster concerning Technology and System Operations, characterized by the salient frequency of terms including "technology," "system," "data," "algorithm," "innovation," and "automation." This lexicon signifies a substantive shift in consumer discourse from a purely experiential or utilitarian engagement with AI toward an active interest in its underlying technical foundations. The recurring invocation of these operational terms suggests a collective desire to comprehend the mechanistic logic of AI systems, countering perceptions of them as opaque or mystical "black boxes." This represents a move beyond passive acceptance, indicating a burgeoning public demand for operational transparency and a more granular understanding of how data fuels algorithmic processes and automated outcomes. This engagement can be interpreted as an emergent form of algorithmic literacy (Kellogg et al. 2020), wherein consumers are developing a functional vocabulary and conceptual framework to critically interrogate the technological infrastructure shaping their retail experiences. Consequently, technical comprehension is becoming a key factor in the assessment of AI's legitimacy, reliability, and appropriate application within the commercial sphere.

In synthesizing the thematic analysis, this investigation into consumer discourse on AI in retail has elucidated six principal themes that collectively map the contours of public perception and negotiation. First, the Interaction Between People and AI highlights a prevailing view of AI as an augmentative tool, with an emphasis on preserving human oversight. Second, significant Employment and the Future of Work anxieties reveal deep-seated concerns about labor displacement and economic restructuring. Third, recognition of AI as a driver of Business and Market Evolution underscores an awareness of its strategic, competitive implications. Fourth, the centrality of the Quality of Customer Experience demonstrates that AI's value is ultimately judged by its ability to enhance service interactions. Fifth, prominent Ethical and Privacy Issues point to critical demands for accountability and transparency in data practices. Finally, the theme of Technology and System Operations indicates a growing consumer curiosity and algorithmic literacy (Kellogg et al. 2020), reflecting a desire to demystify the technical workings of AI systems. Together, these themes illustrate a complex and multifaceted public engagement with AI, where pragmatic assessments of utility and efficiency are deeply intertwined with socio-economic anxieties, ethical scrutiny, and an evolving understanding of the technology itself.

The identified themes are not isolated but are dynamically interconnected, forming a complex web of consumer perception. For instance, apprehensions regarding

Employment and the Future of Work directly influence evaluations of the Quality of Customer Experience; interactions with automated systems may be viewed less favorably if perceived as contributing to job displacement. Similarly, concerns over Ethical and Privacy Issues and the desire for Technology and System Operations transparency are fundamentally linked, as trust in algorithmic systems is contingent upon understanding and approving of their underlying data practices. This trust, in turn, mediates acceptance of AI as a tool for Business and Market Evolution. Furthermore, the core Interaction Between People and AI theme acts as the experiential nexus where all other themes, ethical scrutiny, employment anxiety, technical curiosity, and experiential evaluation, converge in the actual moment of consumer engagement. These interrelations suggest that public acceptance of AI in retail is a holistic calculation, where technical capability, economic impact, ethical governance, and human-centric design are weighed simultaneously.

3.5 *Academic Summary: Areas for Further Research*

Building upon these findings, several critical avenues for further research emerge. Methodologically, a mixed-methods approach combining computational text analysis with qualitative interviews or structured surveys would be invaluable to capture the nuanced motivations and contextual reasoning behind the discursive patterns identified here (Creswell and Plano Clark 2017). Longitudinal studies are necessary to track how public perceptions evolve as AI technologies mature and become further embedded in retail ecosystems, applying models of technology acceptance and diffusion (Venkatesh et al. 2016). Comparative research is also essential to determine the influence of cultural dimensions, demographic variables, and sector-specific contexts on attitude formation, testing whether the themes identified hold across different national cultures, age cohorts, or retail sub-sectors (Hofstede 2001). Finally, experimental or design-based research could proactively test interventions, such as enhanced system transparency or user education programs, to evaluate their efficacy in bridging identified "algorithmic accountability gaps" and building public trust in AI-driven retail environments (Eslami et al. 2018). Pursuing these lines of inquiry will deepen the scholarly and practical understanding of the human factors critical to the responsible and successful integration of AI in commerce.

In conclusion, this thematic analysis elucidates a complex and dialectical public negotiation of artificial intelligence within the contemporary retail landscape. The discursive data reveal a collective consciousness engaged in a strategic appraisal, simultaneously acknowledging AI's instrumental utility in driving efficiency and innovation while critically interrogating its profound socio-economic, ethical, and phenomenological ramifications. Consumers emerge not as passive adopters but as active sense-makers, deliberating on AI's transformative influence on fundamental structures: the ontology of work, the dynamics of market competition, the fabric of customer experience, and the architecture of privacy. The persistent lexical salience of "human" and "person" across the thematic framework is particularly revelatory, signaling a powerful, collective insistence on centering human agency, values, and oversight within increasingly algorithmic systems. Consequently, for both corporate strategists and regulatory bodies, this research underscores a critical imperative: the trajectory of AI integration in retail must be guided by a framework that transcends mere technological optimization and economic gain. Its long-term viability and legitimacy are contingent upon the adoption of rigorously human-centered, ethically transparent, and socially responsible paradigms, ensuring alignment with the nuanced spectrum of public expectation and normative concern.

Key Enhancements:

Precision of Language: Uses terms like "elucidates," "dialectical," "instrumental utility," and "phenomenological ramifications" for academic depth.

Stronger Framing: Positions the conclusion as revealing a "dialectical" negotiation and describes consumers as debating the "ontology of work" and "architecture of privacy."

Academic Flow: Connects ideas with phrases like "Consequently," and "its long-term viability and legitimacy are contingent upon," which are common in academic conclusions to state implications firmly.

This version maintains your original insight and structure while aligning closely with the stylistic conventions of published social science literature.

4 Conclusion

This study has undertaken a critical discourse analysis to navigate the profound algorithmic shift reshaping digital marketing agencies. By examining the intersection of AI adoption, managerial sensemaking, and strategic adaptation, the research moves beyond a purely technological or functional assessment to reveal the complex socio-cultural and interpretive processes that mediate AI's impact on agency performance and growth.

The analysis confirms that AI adoption represents a dual-force paradigm for digital agencies. On one hand, it drives undeniable operational efficiency through the automation of repetitive tasks (e.g., reporting, A/B testing, basic content production), compressing timelines and reducing the cost-of-service delivery. On the other hand, it enables transformative strategic effectiveness by augmenting human intelligence for hyper-personalization, predictive analytics, and real-time optimization, thereby enhancing client outcomes and value proposition. The findings suggest that the most significant performance implication lies not in choosing between these streams, but in leveraging their synergy: efficiency gains liberate human capital to focus on the higher-order strategic and creative work that drives effectiveness.

However, the core contribution of this research is its illumination of the interpretive and discursive labor required to harness this potential. The discourse analysis reveals that AI integration is not a straightforward technical implementation but a continuous process of managerial sensemaking. Agency leaders and practitioners must actively interpret, legitimize, and translate AI's value, both for internal stakeholders navigating cultural resistance and capability gaps and for external clients managing expectations and anxieties. The tension between the consistent, brand-safe output of AI systems and the divergent, innovative potential of human creativity emerged as a central strategic dilemma. This necessitates a contingent, hybrid approach where the choice of generation mode (AI-automated, human-led, or collaborative) is strategically matched to content type and campaign goal.

Furthermore, the study underscores that sustainable strategic growth in this new landscape requires more than tool adoption. It demands the development of dynamic capabilities, where AI functions as the engine for continuously sensing market shifts, seizing opportunities, and reconfiguring agency resources. This involves a fundamental business model evolution, from selling billable hours for manual execution to offering premium services as strategic orchestrators, AI consultants, and guardians of algorithmic performance. Success hinges on building a VRIN (valuable, rare, imperfectly imitable, non-substitutable) resource bundle combining proprietary data, technical talent, and a culture of adaptive learning.

In conclusion, the algorithmic shift is fundamentally rewiring the DNA of digital agencies. The path to performance and growth is no longer linear but navigational, requiring agencies to become bi-modal organizations. They must master the disciplined, data-driven science of AI optimization while simultaneously cultivating the intuitive, human-centric art of strategy, creativity, and ethical stewardship. The future belongs not to agencies that merely use AI, but to those that can strategically discourse with it, interpreting its outputs, integrating its logic into their culture, and continuously adapting their human-machine collaboration to forge new, defensible, and value-creating roles in the marketing ecosystem. This research provides a foundational framework for understanding that journey, highlighting that the ultimate competitive advantage lies in the metaknowledge of when, how, and why to blend artificial intelligence with irreplaceable human insight.

References

Acemoglu, D., Restrepo, P.: Robots and jobs: Evidence from US labor markets. J. Polit. Econ. **128**(6), 2188–2244 (2020)

Barney, J.: Firm resources and sustained competitive advantage. J. Manag. **17**(1), 99–120 (1991)

Borges, A.F., Laurindo, F.J., Spínola, M.M., Gonçalves, R.F., Mattos, C.A.: The strategic use of artificial intelligence in the digital era: systematic literature review and future research directions. Int. J. Inf. Manag. **57**, 102225 (2021)

Bretthauer, K.M., Mahar, S., Venkatadri, U.: Service unbundling and rebundling: a dynamic approach to digital platform strategy. Decis. Sci. **48**(4), 775–807 (2017)

Brynjolfsson, E., McAfee, A.: The Second Machine Age: Work, Progress, and Prosperity in a Time of Brilliant Technologies. W. W. Norton & Company (2014)

Creswell, J.W., Plano Clark, V.L.: Designing and Conducting Mixed Methods Research, 3rd edn. SAGE Publications (2017)

Davenport, T., Guha, A., Grewal, D., Bressgott, T.: How artificial intelligence will change the future of marketing. J. Acad. Mark. Sci. **48**(1), 24–42 (2020)
Eslami, M., Rickman, A., Vaccaro, K., Aleyasen, A., Vuong, A., Karahalios, K., Hamilton, K., Sandvig, C.: "I always assumed that I wasn't really that close to [her]": reasoning about invisible algorithms in news feeds. CHI Conf. Hum. Factors Comput. Syst., 153–165 (2018)
Hofacker, C.F., Malthouse, E.C., Sultan, F.: Big data and consumer behavior: imminent opportunities. J. Consum. Mark. **33**(2), 89–97 (2016)
Hofstede, G.: Culture's Consequences: Comparing Values, Behaviors, Institutions and Organizations across Nations, 2nd edn. SAGE Publications (2001)
Huang, M.H., Rust, R.T.: Artificial intelligence in service. J. Serv. Res. **21**(2), 155–172 (2018)
Huang, M.H., Rust, R.T.: A strategic framework for artificial intelligence in marketing. J. Acad. Mark. Sci. **49**(1), 30–50 (2021)
Kaplan, A., Haenlein, M.: Siri, Siri, in my hand: Who's the fairest in the land? On the interpretations, illustrations, and implications of artificial intelligence. Bus. Horiz. **62**(1), 15–25 (2019)
Kellogg, K.C., Valentine, M.A., Christin, A.: Algorithms at work: the new contested terrain of control. Acad. Manag. Ann. **14**(1), 366–410 (2020)
Kietzmann, J., Paschen, J., Treen, E.: Artificial intelligence in advertising: how marketers can leverage artificial intelligence along the consumer journey. J. Advert. Res. **58**(3), 263–267 (2018)
Kumar, V.: Transformative marketing: the next 20 years. J. Mark. **82**(4), 1–12 (2018)
Kumar, V., Dixit, A., Javalgi, R.G., Dass, M.: Research framework, strategies, and applications of intelligent agent technologies (IATs) in marketing. J. Acad. Mark. Sci. **44**(1), 24–45 (2016)
Kumar, V., Rajan, B., Venkatesan, R., Lecinski, J.: Understanding the role of artificial intelligence in personalized engagement marketing. Calif. Manag. Rev. **61**(4), 135–155 (2019)
Lambrecht, A., Tucker, C.: Algorithmic bias? An empirical study of apparent gender-based discrimination in the display of STEM career ads. Manag. Sci. **65**(7), 2966–2981 (2019)
Lemon, K.N., Verhoef, P.C.: Understanding customer experience throughout the customer journey. J. Mark. **80**(6), 69–96 (2016)
Mittelstadt, B.D., Allo, P., Taddeo, M., Wachter, S., Floridi, L.: The ethics of algorithms: mapping the debate. Big Data Soc. **3**(2), 1–21 (2016)
Pine, B.J., Gilmore, J.H.: Welcome to the experience economy. Harv. Bus. Rev. **76**(4), 97–105 (1998)
Raffaelli, R., Glynn, M.A., Tushman, M.: Frame flexibility: the role of cognitive and emotional framing in innovation adoption by incumbent firms. Strateg. Manag. J. **40**(7), 1013–1039 (2019)
Syam, N., Sharma, A.: Waiting for a sales renaissance in the fourth industrial revolution: machine learning and artificial intelligence in sales research and practice. Ind. Mark. Manag. **69**, 135–146 (2018)
Teece, D.J.: Explicating dynamic capabilities: the nature and microfoundations of (sustainable) enterprise performance. Strateg. Manag. J. **28**(13), 1319–1350 (2007)
Teece, D.J., Pisano, G., Shuen, A.: Dynamic capabilities and strategic management. Strateg. Manag. J. **18**(7), 509–533 (1997)
Venkatesh, V., Thong, J.Y., Xu, X.: Unified theory of acceptance and use of technology: a synthesis and the road ahead. J. Assoc. Inf. Syst. **17**(5), 328–376 (2016)
Verhoef, P.C., Broekhuizen, T., Bart, Y., Bhattacharya, A., Qi Dong, J., Fabian, N., Haenlein, M.: Digital transformation: a multidisciplinary reflection and research agenda. J. Bus. Res. **122**, 889–901 (2021)
Verma, S., Sharma, R., Deb, S., Maitra, D.: Artificial intelligence in marketing: systematic review and future research direction. Int. J. Inf. Manag. Data Insights. **1**(1), 100002 (2022)

Predictive Technologies and Fraud Risk Anticipation: Evidence from Big Four Auditors in Morocco

Abdelmounim Bouziane

Abstract The increasing digitalization of business processes has intensified interest in predictive technologies as tools for enhancing auditors' ability to anticipate financial fraud risk. While prior research highlights the potential of advanced analytics and artificial intelligence in auditing, empirical evidence on their effective use in external audit settings—particularly in developing economies—remains limited. This study examines how predictive technologies contribute to auditors' anticipation of financial fraud risk within Big Four audit firms operating in Morocco. Using a quantitative, cross-sectional research design, data were collected through a structured questionnaire administered to 82 external auditors. Principal Component Analysis was employed to identify the latent dimensions structuring auditors' perceptions of predictive technologies. The results reveal three key components: (i) predictive technology performance and audit quality enhancement, (ii) organizational technological maturity and governance, and (iii) user acceptance and behavioral readiness. The findings indicate that predictive technologies enhance fraud risk anticipation primarily when software performance, output quality, and prediction reliability are high and supported by stable systems and adequate IT governance. Organizational readiness, including technology maturity and formal IT oversight, further shapes effective integration. Additionally, auditors' acceptance, training, and resistance to change significantly influence the extent to which predictive tools are meaningfully used in audit work. This study contributes to the auditing literature by providing empirical evidence from a developing-country context and by demonstrating that the value of predictive technologies in fraud risk anticipation depends on the interaction between technical, organizational, and human factors. The findings offer practical insights for audit firms seeking to strengthen fraud-related risk assessment through advanced analytics while preserving audit quality and professional judgment.

Keywords Artificial intelligence · Financial audit · Predictive technologies

---

A. Bouziane (✉)
Chouaib Doukkali University & Urban Agency of Casablanca, Casablanca, Morocco
e-mail: bouzianeabdelmounim@gmail.com; bouziane.a@ucd.ac.ma

A. Bouziane and K. Bouziane (eds.), *AI, Management, and Culture: Shaping Organizations in the Digital Age*, Studies in Computational Intelligence 1291,
https://doi.org/10.1007/978-3-032-27882-1_10

1 Introduction

The growing complexity of business operations and the increasing volume of digital transactions have intensified concerns about financial fraud and the effectiveness of traditional audit approaches. High-profile corporate scandals and persistent fraud cases continue to challenge the credibility of financial reporting and reinforce the central role of external auditors in protecting stakeholders' interests. Within this context, the ability to anticipate fraud risk—rather than merely detect misstatements ex post—has become a critical objective of contemporary auditing.

At the same time, the audit profession is undergoing a profound digital transformation. Advances in information technologies, data analytics, and artificial intelligence have reshaped how audit evidence is generated, analyzed, and evaluated. Audit firms increasingly rely on Computer-Assisted Audit Techniques (CAATs) and, more recently, on predictive technologies capable of analyzing large datasets to identify anomalous patterns and forecast areas of elevated fraud risk. These tools promise earlier risk identification, broader coverage of transactions, and more efficient allocation of audit effort, potentially enhancing audit quality and professional judgment.

Despite their potential, the integration of predictive technologies into audit practice raises important questions. Prior research shows that technology adoption alone does not guarantee improvements in audit quality or fraud detection. The effectiveness of digital tools depends on multiple factors, including software performance, output reliability, organizational infrastructure, governance mechanisms, and auditors' acceptance, skills, and resistance to change. While large international audit networks, such as the Big Four, may possess superior technological resources and standardized methodologies, challenges related to implementation, trust in outputs, and human–technology interaction remain relevant.

Existing literature on audit technology has largely focused on CAAT adoption, internal audit contexts, or developed economies. Empirical evidence on predictive technologies used by external auditors, particularly in developing-country settings, remains limited. Moreover, few studies simultaneously examine technical, organizational, and behavioral dimensions to understand how predictive tools contribute to fraud risk anticipation. This gap is particularly important in emerging markets, where institutional environments, regulatory pressures, and technological maturity may shape auditors' experiences differently from those documented in more mature economies.

Against this backdrop, this study investigates auditors' perceptions of predictive technologies and their role in anticipating financial fraud risk within Big Four audit firms operating in Morocco. By adopting a quantitative, perception-based approach and applying Principal Component Analysis, the study seeks to identify the underlying dimensions that structure auditors' views regarding the use, performance, and effectiveness of predictive technologies in audit practice.

Research Question

To address this objective, the study is guided by the following research question:

How do predictive technologies contribute to auditors' anticipation of financial fraud risk, and what technical, organizational, and behavioral factors shape their effective use within Big Four audit firms in Morocco?

By answering this question, the study contributes to the auditing and information systems literature by providing empirical insights into the conditions under which predictive technologies enhance fraud risk anticipation. It also offers practical implications for audit firms seeking to leverage advanced analytics while maintaining high standards of audit quality and professional judgment.

2 Literature Review

2.1 Audit Quality and the Changing Nature of Assurance Work

Audit quality is widely viewed as a cornerstone of corporate governance because it enhances the credibility of financial reporting and reduces information asymmetry between management and stakeholders (Achouri et al. 2022). Early conceptualizations describe audit quality as the likelihood that auditors both detect material misstatements and report them, which places auditor competence and independence at the center of the audit quality discussion (DeAngelo 1981). Subsequent research has continued to emphasize these fundamental drivers while also acknowledging that audit quality is shaped by broader contextual influences, such as organizational environment and behavioral factors within audit teams (Gaddour 2016), and even cultural dimensions that affect professional judgments and practices (Bouziane et al. 2023a, b).

As auditing becomes more technology-intensive, the meaning and determinants of audit quality evolve. Digitalization alters audit evidence, the nature of tests performed, and the speed at which risks must be assessed. Consequently, audit quality is increasingly linked to how effectively auditors can use technological tools to produce timely and relevant evidence without compromising professional judgment.

2.2 Digital Transformation in Auditing: From ICT to CAATs

Information and Communication Technologies have reshaped accounting systems and audit work methods, enabling audit firms to improve efficiency in communication, analysis, documentation, and archiving (Mutchler et al. 1997). Empirical and conceptual work suggests that technology adoption can strengthen audit procedures and enhance the resulting quality of assurance (Bierstaker et al. 2001). However, technology also creates new professional pressures: auditors face increased expectations related to both technical expertise and audit value-added, and may also experience organizational and social impacts from intensified monitoring, standardization, and workload acceleration (Gwilliam et al. 2000; Matthews 2006; Robson et al. 2007).

Within this technological shift, Computer-Assisted Audit Techniques (CAATs) have emerged as a major category of tools supporting audit testing and audit planning. Unlike traditional approaches that depend heavily on sampling, CAATs can facilitate testing of broader—sometimes complete—data populations, potentially strengthening the evidence base and improving assurance confidence (Singleton et al. 2006). CAATs can also accelerate testing, allowing auditors to perform faster risk assessments and allocate audit effort more efficiently, which is especially relevant under rising workload constraints and stakeholder expectations (Ghosh and Pawlewicz 2009). Additionally, CAATs are often positioned as effective tools for identifying anomalies indicative of errors or fraud, such as duplicates, missing transactions, or unusual patterns (Coderre 1999), supporting the view that software-based testing can enhance audit effectiveness if implemented appropriately (Braun and Davis 2003).

Evidence from developing contexts indicates that CAAT usage is positively associated with audit quality, but also that effectiveness depends on resources and capabilities. For example, findings from Nigeria suggest that while CAATs are used and are linked to audit quality, smaller local firms may struggle to apply them effectively compared with large international networks that can invest in infrastructure and specialized skills (Omonuk and Oni 2015). Related research similarly emphasizes that CAATs can improve efficiency and effectiveness, but the realized benefits depend on the firm's investment capacity and auditors' expertise (Eulerich et al. 2023; Nasrah and Muda 2023).

2.3 From CAATs to Predictive Technologies: Anticipating Fraud Risk

The move from CAATs toward predictive technologies represents a shift in focus: from testing historical transactions to anticipating risk using advanced analytical methods. Predictive tools can be viewed as an extension of computer-assisted auditing in which auditors rely on performance-driven software outputs to detect abnormal patterns and to forecast where fraud risk is more likely to arise. As organizations digitize their processes and generate high-volume transaction data, manual auditing becomes less feasible, strengthening the rationale for data-driven methods (Meiryani and Susanto 2018).

Within fraud risk work specifically, predictive technologies may provide earlier warning signals and more targeted audit attention by identifying clusters of high-risk transactions or entities. This aligns with the broader argument that technology enhances auditors' capacity to move from routine verification toward higher-value judgment, so long as the tools produce reliable outputs and are embedded properly in audit methodology.

2.4 *Technology Acceptance, Competence, and Organizational Readiness*

Even when predictive tools are technically powerful, their contribution to audit quality and fraud risk anticipation is limited if auditors do not accept or effectively use them. Research on technology adoption in auditing highlights the importance of perceived usefulness and ease of use (Pedrosa et al. 2019). This implies that auditors are more likely to integrate digital tools into audit planning and testing when they believe the tools improve task performance and are feasible to use in real engagements.

However, adoption is not purely attitudinal—it is also capability-based. Auditors may recognize the importance of technology while still reporting inadequate practical knowledge, suggesting a skills gap that reduces the effectiveness of implementation (Al-Duwaila and Al-Mutairi 2017). This gap reinforces the need for structured training, not only in professional standards but also in the operational use of audit technologies and interpretation of outputs. Consistent with this view, auditors' competence is repeatedly emphasized as a key factor in optimizing the use and performance of CAATs (Purnamasari and Hartanto 2022). The same logic applies to predictive technologies: sophisticated tools require users who can interpret results, challenge anomalies, and integrate findings into audit judgment.

Research in internal audit settings further supports the idea that adoption depends on resources, awareness of benefits, training, and infrastructure. For example, evidence from Jordan indicates that decision-makers can increase CAAT use by communicating benefits, investing in technical infrastructure, and strengthening users' skills through training programs (Al-Hiyari et al. 2019). Although internal and external audits differ in mandate and stakeholders, these findings reinforce a broader conclusion: technology-driven audit quality depends on readiness at both the individual (skills, confidence) and organizational (systems, processes, support) levels.

2.5 *AI-Enabled Tools and Audit Quality: Performance, Reliability, and Governance Issues*

Recent literature on AI in auditing—especially in internal audit—highlights both significant opportunities and important constraints. Intelligent Process Automation is proposed as a means of shifting auditors away from repetitive tasks so they can concentrate on judgment-intensive work, though researchers also emphasize the need for empirical validation of impacts on quality and efficiency (Zhang 2019). Evidence suggests that audit functions adopting AI may experience increased demand for auditors with stronger interpersonal and judgment-related skills, while quality may improve without necessarily displacing jobs (Law and Shen 2020). Investments in

AI-related talent and capabilities are also associated with higher-quality outputs and fewer inefficiencies (Fedyk et al. 2022).

Work focused on ChatGPT and large language models highlights practical gains—speed, support for standardized tasks, and assistance for junior staff—alongside concerns about accuracy, reliability, and ethical risks (Fotoh and Mugwira 2023). Case-based evidence indicates time reductions for some audit tasks and potential improvements in workflow, while still requiring careful risk management regarding biased outputs, confidentiality, and legal exposure (Emett et al. 2023). Research on "AI co-piloted auditing" further frames AI as a collaborator that can support complex audit tasks when guided through structured prompting and decomposition of tasks, but with the implication that strong governance and human oversight remain essential (Gu et al. 2023). Similarly, practical discussions emphasize that successful integration requires standardized processes, motivated staff, high data quality, and robust IT infrastructure, while managing risks such as privacy, security, and accountability (Eulerich and Wood 2023).

Taken together, these studies suggest that the value of advanced tools in auditing is likely to depend less on "AI presence" alone and more on tool performance, output quality, reliability, and the governance framework around their use. This directly supports the logic of examining variables such as software performance, quality of results, and prediction reliability when studying fraud risk anticipation.

2.6 Linking Prior Research to This Study's Framework

The literature supports a progression from foundational audit quality drivers (competence, independence) (DeAngelo 1981) toward technology-enabled audit effectiveness (Bierstaker et al. 2001; Singleton et al. 2006) and then toward advanced analytics and AI-like applications that may enhance risk assessment and fraud-related work (Zhang 2019; Fedyk et al. 2022; Emett et al. 2023). Across these streams, a recurring pattern emerges:

Perceived usefulness and feasibility matter for adoption and day-to-day integration of tools (Pedrosa et al. 2019), and these perceptions can be constrained by limited user competence (Al-Duwaila and Al-Mutairi 2017; Purnamasari and Hartanto 2022).

Organizational context and resources shape whether technology is used effectively, particularly in developing-country environments (Omonuk and Oni 2015; Al-Hiyari et al. 2019).

Performance, quality of outputs, and reliability are central for translating tool use into improved audit outcomes; without trustworthy results, auditors may hesitate to rely on outputs for critical judgments (Fotoh and Mugwira 2023; Eulerich and Wood 2023).

Therefore, examining predictive technologies in auditing—especially for fraud risk anticipation—requires focusing on both (i) auditors' acceptance and feasibility perceptions, and (ii) the technical and practical attributes that determine whether outputs can be trusted and integrated into professional judgment. In Big Four settings, where resources and standardized methodologies may be stronger, the decisive differentiators may be software performance and output quality, while usefulness and feasibility remain important but more contingent on auditors' expertise and engagement context.

3 Method

3.1 Research Design

This study adopts a quantitative, cross-sectional research design to examine how predictive technologies contribute to auditors' anticipation of financial fraud risk. Data were collected using a structured questionnaire administered to auditors working in Big Four audit firms in Morocco. The methodological choices align with prior audit-technology research that relies on perception-based measures to capture technology-related determinants in professional settings.

3.2 Population, Sampling, and Participants

The target population consists of external auditors employed by Big Four firms operating in Morocco. A non-probability sampling approach (convenience and professional network-based recruitment) was used due to the practical constraints of accessing auditors during engagement periods and the confidentiality norms governing audit practice.

A total of 82 out of 124 (total auditors working in big four firms in Morocco) usable responses were obtained. Respondents included auditors with varying levels of experience and responsibilities (e.g., junior auditors, seniors, and managers).

3.3 Data Collection Procedure

Data were collected through a questionnaire distributed electronically. Participants were informed of the academic purpose of the study, the voluntary nature of participation, and the confidentiality of responses. To reduce social desirability bias and encourage honest responses, the questionnaire did not request identifying information (e.g., name, client identity, engagement details).

Table 1 Research variables

Variables—questionnaire
Performance of software used
Perceived usefulness
Perceived ease
Quality of results
Prediction reliability
Resistance to change
Audit quality
Training programs offered to auditors
Technology maturity of the firm
Use of software
IT assistance
Frequency of IT audit
Frequency of breakdowns

3.4 Measurement of Variables and Instrument Structure

The questionnaire was developed based on constructs widely used in the auditing and information systems literature. The research variables were as follows (Table 1).

Items were measured using Likert-type scales (e.g., 1 = strongly disagree to 5 = strongly agree). The questionnaire also included descriptive items to capture respondent characteristics (e.g., years of experience, sexe, position).

3.5 Data Preparation and Screening

Prior to analysis, the dataset was screened to ensure completeness and consistency. Responses were checked for missing values, out-of-range entries, and patterns suggesting non-engaged completion. Only fully usable questionnaires were retained for statistical analysis.

3.6 Data Analysis Strategy: Principal Component Analysis (PCA)

To identify the latent dimensions structuring auditors' perceptions of predictive technologies, the study employed Principal Component Analysis using SPSS. PCA was selected because the research objective is exploratory/dimensional: to reduce a set of correlated observed items into a smaller number of interpretable components representing the key underlying factors.

Participation was voluntary and based on informed consent. Respondents were assured that their answers would remain confidential and would be used strictly for academic purposes. No personal identifiers or sensitive client-specific information were collected, minimizing risks related to professional confidentiality.

4 Results

Before conducting Principal Component Analysis, the suitability of the data was assessed using the **Kaiser–Meyer–Olkin (KMO)** measure of sampling adequacy and **Bartlett's Test of Sphericity**. The results indicate that the correlation matrix is factorable and the PCA is appropriate (Table 2).

PCA was conducted in SPSS on the items measuring perceived usefulness, perceived feasibility, software performance, quality of results, prediction reliability and other variables. Components were retained based on **eigenvalues > 1** and inspection of the scree plot. Table 3 presents the Component Matrix.

The Principal Component Analysis (PCA) resulted in the extraction of three components, which together structure auditors' perceptions of predictive technologies and their role in anticipating financial fraud risk. Interpretation of the components is based on variables with the highest factor loadings (generally $\geq |0.50|$), indicating strong contributions to each component.

Component 1: Perceived Audit Quality and Predictive Technology Effectiveness

Component 1 is primarily defined by strong positive loadings on variables related to audit quality and the technical performance of predictive tools. High loadings are observed for:

- Audit Quality (0.792)
- Quality of results (0.654)
- Prediction reliability (0.649)
- Performance of software used (0.576)
- IT assistance (0.502)
- Frequency of IT audit (0.509)

In contrast, Frequency of breakdowns loads negatively on this component (−0.572), indicating that system instability undermines perceptions of effectiveness and audit quality.

Table 2 Reports the KMO value and Bartlett's test statistics

KMO	0.839
Sig. of Bartlett	0.000

Table 3 Component matrix

	Component 1	Component 2	Component 3
Sexe	0.301	0.503	0.411
Experience	0.204	0.500	0.205
Position	0.441	0.544	0.143
Performance of software used	0.576	0.331	0.204
Perceived usefulness	0.234	0.113	0.640
Perceived ease	0.231	0.212	0.532
Quality of results	0.654	− 0.365	0.303
Prediction reliability	0.649	0.108	0.410
Resistance to change	0.401	0.150	0.566
Audit quality	0.792	0.409	0.206
Training programs offered to auditors	− 0.408	0.398	0.501
Technology maturity of the firm	0.308	0.502	0.400
Use of software	0.208	− 0.185	0.521
IT assistance	0.502	0.175	0.321
Frequency of IT audit	0.509	0.533	0.400
Frequency of breakdowns	− 0.572	− 0.165	0.070

This component reflects auditors' assessment of how robust, reliable, and well-supported predictive technologies enhance audit quality and fraud risk anticipation. When predictive tools perform well, produce high-quality outputs, and are supported by adequate IT infrastructure, auditors perceive a stronger contribution to fraud detection and audit effectiveness.

Component 2: Organizational and Structural Technological Environment

Component 2 is characterized by variables related to the organizational context and firm-level technological infrastructure, including:

- Frequency of IT audit (0.533)
- Position (0.544)
- Technology maturity of the firm (0.502)
- Experience (0.500)
- Moderate loadings for training programs offered to auditors (0.398) and audit quality (0.409) suggest that organizational support mechanisms contribute indirectly to this dimension.

This component captures the institutional readiness of audit firms, where higher hierarchical positions, greater experience, mature technological environments, and formal IT governance practices shape auditors' perceptions of predictive technologies.

Component 3: User Acceptance, Skills, and Behavioral Factors

Component 3 shows strong loadings on variables related to individual acceptance, usage, and behavioral responses toward predictive technologies:

- Perceived usefulness (0.640)
- Resistance to change (0.566)
- Perceived ease of use (0.532)
- Use of software (0.521)
- Training programs offered to auditors (0.501)

This component reflects auditors' cognitive and behavioral engagement with predictive technologies. Higher perceived usefulness and ease of use, supported by training, encourage adoption, while resistance to change remains a critical behavioral barrier.

Overall Interpretation and Link to Fraud Risk Anticipation

Taken together, the three components highlight that fraud risk anticipation through predictive technologies depends on:

Technical performance and audit quality outcomes (Component 1),

Firm-level technological maturity and governance structures (Component 2), and.

Auditors' acceptance, training, and behavioral readiness (Component 3).

These findings suggest that effective anticipation of fraud risk in Big Four audit firms in Morocco is not solely driven by advanced tools, but by the interaction between technology performance, organizational support, and human factors.

5 Discussion

This study examined how predictive technologies contribute to auditors' anticipation of financial fraud risk in Big Four firms operating in Morocco. Using Principal Component Analysis, three distinct but complementary dimensions emerged: (i) predictive technology performance and audit quality enhancement, (ii) organizational technological maturity and governance, and (iii) user acceptance and behavioral readiness. Together, these dimensions provide empirical support for prior research while extending existing knowledge to the context of predictive technologies and fraud risk anticipation in a developing-country setting.

Predictive technology performance and audit quality enhancement

The first component highlights the central role of software performance, quality of results, prediction reliability, IT support, and audit quality, with a negative association for system breakdowns. This finding aligns strongly with prior literature emphasizing that audit quality increasingly depends on the technical robustness and reliability of digital tools, rather than on technology adoption alone.

Consistent with DeAngelo's (1981) foundational view of audit quality as the ability to detect and report misstatements, predictive technologies appear to enhance auditors' detection capacity when outputs are reliable and systems are stable. The strong loading of audit quality on this component supports earlier evidence that technology-enabled audits can improve assurance outcomes when tools are effectively implemented (Bierstaker et al. 2001; Singleton et al. 2006). Moreover, the importance of prediction reliability and output quality echoes recent research on AI-enabled auditing, which stresses that auditors' willingness to rely on advanced tools depends on the perceived trustworthiness of results (Fotoh and Mugwira 2023; Eulerich and Wood 2023).

The negative loading of system breakdown frequency further reinforces concerns raised in prior studies regarding technological fragility. While predictive tools can enhance fraud risk anticipation, unstable systems may undermine confidence and discourage reliance on automated outputs, supporting arguments that performance risks and operational failures remain critical barriers to effective digital auditing (Emett et al. 2023). Overall, this component confirms that fraud risk anticipation is strengthened not by technology presence per se, but by reliable, well-supported, and high-performing systems.

Organizational technological maturity and governance

The second component captures variables related to technology maturity, IT audit frequency, hierarchical position, and experience, pointing to the importance of organizational context in shaping perceptions of predictive technologies. This finding is consistent with earlier studies showing that technology effectiveness in auditing is strongly influenced by firm-level resources, governance structures, and standardized processes (Omonuk and Oni 2015; Al-Hiyari et al. 2019).

The prominence of technology maturity and IT audit frequency suggests that predictive technologies are more likely to support fraud risk anticipation when embedded in a controlled and monitored technological environment. This aligns with research emphasizing governance mechanisms—such as periodic IT audits and formal controls—as prerequisites for trustworthy digital audit outputs (Eulerich and Wood 2023). The association with experience and position further suggests that senior auditors and managers may be more attuned to the strategic role of predictive technologies, reflecting prior findings that technology-driven audit transformation is often guided by higher-level professionals (Gwilliam et al. 2000; Robson et al. 2007).

Importantly, this component supports the argument that large international audit networks, such as the Big Four, possess structural advantages in adopting advanced technologies due to their investment capacity and methodological standardization (Omonuk and Oni 2015). In the Moroccan context, this organizational readiness appears to be a key enabler for integrating predictive tools into fraud risk assessment practices.

User acceptance and behavioral readiness

The third component reflects perceived usefulness, perceived ease of use, resistance to change, training programs, and actual software use, highlighting the behavioral and competence-related dimensions of predictive technology adoption. This result closely aligns with technology acceptance research in auditing, particularly studies emphasizing perceived usefulness and ease of use as critical determinants of adoption (Pedrosa et al. 2019).

The significant loading of resistance to change indicates that, even in well-resourced Big Four environments, behavioral barriers remain relevant. This supports prior evidence that auditors may hesitate to fully embrace advanced tools due to uncertainty, lack of confidence, or disruption to established work routines (Al-Duwaila and Al-Mutairi 2017). The importance of training programs reinforces the view that effective use of predictive technologies requires not only positive attitudes but also practical skills and interpretive competence (Purnamasari and Hartanto 2022).

These findings are consistent with recent discussions on AI-assisted auditing, which emphasize that advanced tools function best as complements to human judgment rather than substitutes, and that auditors must be trained to critically interpret outputs and manage associated risks (Gu et al. 2023; Fotoh and Mugwira 2023). Thus, user acceptance and readiness remain essential conditions for translating predictive technology capabilities into meaningful improvements in fraud risk anticipation.

Integrated interpretation and contribution to prior research

Taken together, the results reinforce a multidimensional view of predictive technology use in auditing. Consistent with prior literature, the findings suggest that effective fraud risk anticipation emerges from the interaction between technical performance, organizational readiness, and human factors, rather than from any single dimension alone. While Big Four firms may benefit from superior infrastructure and governance, the realization of value from predictive technologies still depends on auditors' skills, acceptance, and confidence in outputs.

This study extends existing research by empirically demonstrating how these dimensions co-exist and structure auditors' perceptions in a developing-country context. In doing so, it supports recent calls for moving beyond binary measures of technology adoption toward more nuanced analyses that consider performance, reliability, and governance alongside user acceptance (Zhang 2019; Fedyk et al. 2022).

Overall, the findings suggest that predictive technologies can enhance fraud risk anticipation in external auditing, but only when supported by reliable systems, mature organizational frameworks, and adequately trained and engaged auditors. This integrated perspective contributes to the growing literature on technology-enabled assurance and offers practical insights for audit firms seeking to strengthen fraud-related risk assessment through advanced analytics.

6 Recommendations

Based on the study's findings—showing that quality of results and software performance are the strongest drivers of fraud risk anticipation, while usefulness, feasibility, and prediction reliability remain important under certain conditions—the following recommendations are proposed for audit firms (especially Big Four in Morocco), regulators/professional bodies, and technology providers.

(1) Recommendations for audit firms (Big Four and large networks).

•Prioritize output quality in tool selection and renewal
•Audit firms should adopt or retain predictive tools only when they consistently produce actionable, interpretable, and audit-relevant outputs (e.g., clear risk flags, explainable anomalies, useful prioritization of high-risk areas). Tool evaluation should go beyond feature lists and focus on whether outputs meaningfully improve fraud risk planning and fieldwork.
•Establish performance benchmarks aligned with engagement constraints
•Because software performance strongly influences adoption and effective use, firms should define minimum standards for:
•processing speed (large datasets),
•system stability (low failure rates),
•compatibility with client data formats (ERP exports, ledgers), and
•integration with audit platforms and documentation tools.
•Performance testing should be conducted before deployment and reviewed periodically.
•Build a formal reliability and validation framework for predictive outputs
•To increase trust and reduce decision risk, firms should implement governance procedures such as:
•model/output validation protocols (accuracy checks, false positive/negative tracking),
•periodic recalibration and monitoring,
•documentation of assumptions and limitations, and
•sign-off rules clarifying when predictive results can influence audit planning, sampling, or expanded procedures.
•This strengthens defensibility and reduces algorithm aversion.
•Strengthen auditors' analytics competence through structured training pathways
•Perceived feasibility depends heavily on auditors' capability. Firms should provide continuous training on:
•using predictive tools (hands-on case exercises),
•interpreting model outputs critically,
•linking outputs to fraud risk factors and audit assertions, and
•documenting rationale and professional judgment.
•Training should be tiered (junior–senior–manager) and embedded into performance evaluation.
•Integrate predictive technologies into audit methodology, not as add-ons
•Predictive tools are most effective when embedded into standardized audit workflows (planning → risk assessment → testing → reporting). Firms should develop practical guidance such as:

-where predictive analytics fits in the fraud brainstorming phase,

-how it informs audit program tailoring,
-how exceptions are investigated, and
-how results affect audit evidence evaluation.

•Improve data readiness and client onboarding practices
•Since output quality depends on data quality, audit teams should implement early-stage "data readiness" checks: completeness, consistency, audit trail integrity, and access controls. Where client systems are heterogeneous, firms should use standardized extraction templates and data cleaning protocols to reduce noise and increase predictive value.
•Maintain "human-in-the-loop" controls for professional skepticism
•Predictive outputs should be treated as decision-support, not automatic conclusions. Audit teams should adopt policies requiring:

-professional skepticism checks against alternative explanations,
-review of high-risk flags by experienced staff, and
-documentation of why outputs were accepted or overridden.

Recommendations for regulators and professional bodies (Morocco)

- Develop guidance on the use of predictive analytics in audits
- Professional bodies can issue guidance on: acceptable use cases, documentation expectations, reliability considerations, and ethical requirements (confidentiality, data handling). This supports consistent practice and protects audit credibility.
- Encourage capacity building and certification in audit analytics
- Institutions can promote specialized training modules and certifications in audit technology, fraud analytics, and AI governance to reduce skill gaps and strengthen the profession's readiness.
- Promote standardization in digital financial reporting and access to audit data
- Regulators can encourage greater standardization of digital records and audit trails, which improves the feasibility and reliability of predictive tools across diverse client environments.

Recommendations for technology providers and audit software developers

- Focus on explainability and interpretability features
- Tools should provide clear explanations of risk flags and anomaly drivers (e.g., "why this transaction is risky"), enabling auditors to defend conclusions and improve trust.
- Optimize tools for real audit conditions
- Providers should enhance:

  – processing speed on large ledgers,
  – compatibility with local ERP/accounting systems, and
  – flexible integration with audit documentation platforms.

- Provide reliability metrics and monitoring dashboards

- Built-in reporting on false positives/negatives, drift indicators, and model stability helps firms govern and validate tool outputs over time.

Managerial roadmap (practical implementation plan)
A practical phased approach for audit firms could be:

- Phase 1: Diagnostic — identify fraud-prone cycles, map available data, assess readiness.
- Phase 2: Pilot — deploy tools in a limited number of engagements; track performance, reliability, and user feedback.
- Phase 3: Standardize — integrate into audit methodology; publish internal guidance and templates.
- Phase 4: Scale — extend across teams; establish KPI monitoring (time savings, detection rates, exception resolution).
- Phase 5: Continuous improvement — retrain models, refine thresholds, update training.

7 Conclusion

This study investigated the role of predictive technologies in auditors' anticipation of financial fraud risk, drawing evidence from Big Four audit firms operating in Morocco. By adopting a quantitative, perception-based approach and applying Principal Component Analysis, the research identified three interrelated dimensions shaping auditors' views of predictive technologies: predictive technology performance and audit quality enhancement, organizational technological maturity and governance, and user acceptance and behavioral readiness.

The findings indicate that predictive technologies contribute to fraud risk anticipation primarily when they deliver reliable, high-quality outputs and are supported by stable systems and adequate IT assistance. Audit quality emerged as closely linked to software performance, prediction reliability, and the absence of system breakdowns, reinforcing the idea that advanced tools enhance assurance only when auditors can trust and operationalize their outputs. Beyond technical attributes, the results underscore the importance of organizational context. Technology maturity, formal IT governance mechanisms, and auditors' experience and hierarchical position play a critical role in shaping how predictive tools are perceived and integrated into audit work.

At the individual level, auditors' acceptance of predictive technologies—captured through perceived usefulness, ease of use, resistance to change, and training—remains a key determinant of effective utilization. Even within well-resourced Big Four environments, behavioral and competence-related factors influence whether predictive tools are used meaningfully in fraud risk assessment. This confirms that technological advancement in auditing does not eliminate the central role of professional judgment, but rather increases the need for skilled auditors capable of interpreting and challenging technology-generated outputs.

Overall, this study contributes to the auditing literature by providing empirical evidence from a developing-country context and by demonstrating that fraud risk anticipation through predictive technologies depends on a combination of technical performance, organizational readiness, and human factors. The results suggest that audit firms seeking to strengthen fraud-related risk assessment should focus not only on investing in advanced tools, but also on ensuring robust governance structures, reliable systems, and continuous auditor training.

Despite its contributions, the study is subject to limitations. The use of non-probability sampling and perception-based measures may limit generalizability and introduce subjective bias. Future research could extend this work by incorporating longitudinal designs, objective performance measures, or comparative analyses across different firm sizes or regulatory environments. Further investigation into how predictive technologies are embedded into specific audit phases and decision-making processes would also deepen understanding of their impact on audit quality and fraud risk anticipation.

References

Achouri, S., Ghazlane, I., El Menzhi, K.: La qualité d'audit: pilier de bonne gouvernance et de performance financière des entreprises. Moroc. J. Res. Manag. Mark. **14**(2) (2022)

Al-Duwaila, N., Al-Mutairi, A.: The opinion of auditors towards the importance and knowledge of information technology in Kuwait. Int. J. Bus. Manag. **12**(3), 170–179 (2017)

Al-Hiyari, A., Said, N.A., Hattab, E.: Factors that influence the use of computer assisted audit techniques (CAATs) by internal auditors in Jordan. Acad. Account. Financ. Stud. J. **23** (2019)

Bierstaker, J.L., Burnaby, P., Thibodeau, J.: The impact of information technology on the audit process: an assessment of the state of the art and implications for the future. Manag. Audit. J. **16**(3), 159–164 (2001)

Bouziane, A., Tahri, W., Bouziane, K.: Acceptability and satisfaction with ICT in learning activities for students: the COVID-19 situation in Morocco. In: Tomczyk, L. (ed.) New Media Pedagogy: Research Trends, Methodological Challenges and Successful Implementations (NMP 22). Springer, Cham (2023a)

Bouziane, A., Tahri, W., Bouziane, K.: The role of national cultural dimensions in the quality of legal audit missions. Int. J. Soc. Cult. Lang. **11**(1) (2023)

Braun, R., Davis, H.: Computer-assisted audit tools and techniques: analysis and perspectives. Manag. Audit. J. **18**(9), 725–731 (2003)

Coderre, D.: Computer assisted techniques for fraud detection. CPA J. **69**(8), 57–59 (1999)

DeAngelo, L.E.: Auditor size and audit quality. J. Account. Econ. **3**(3), 183–199 (1981)

Emett, S.A., Eulerich, M., Lipinski, E., Prien, N., Wood, D.A.: Leveraging ChatGPT for enhancing the internal audit process: a real-world example from a large multinational company. Working paper, SSRN (2023)

Eulerich, M., Masli, A., Pickerd, J., Wood, D.: The impact of audit technology on audit task outcomes: evidence for technology-based audit techniques. Contemp. Account. Res. **40**(2), 981–1012 (2023)

Eulerich, M., Wood, D.A.: A demonstration of how ChatGPT can be used in the internal auditing process. Working paper, SSRN (2023)

Fedyk, A., Khimich, T., Hodson, J., Fedyk, N.V.: Is artificial intelligence improving the audit process? Rev. Acc. Stud. **27**(3), 938–985 (2022)

Fotoh, L., Mugwira, T.: The use of ChatGPT in external audits: implications and future research. Working paper, SSRN (2023)
Gaddour, I.: *Contribution à l'étude de la qualité de l'audit: Une approche fondée sur le management des équipes et le comportement des auditeurs*. Doctoral dissertation, HAL Open Science (2016)
Ghosh, A., Pawlewicz, R.: The impact of regulation on auditor fees: Evidence from the Sarbanes-Oxley Act. Audit.: J. Pract. Theory **28**(2) 171–197 (2009)
Gu, H., Schreyer, M., Moffitt, K., Vasarhelyi, M.A.: Artificial intelligence co-piloted auditing. Working paper, SSRN (2023)
Gwilliam, D., Macve, R., Meeks, G.: Principals and agents in crisis: reforms of accounting and audit at Lloyd's, 1982–1986. Account. Hist. **5**(2), 61–92 (2000)
Law, K., Shen, M.: How does artificial intelligence shape audit firms? Nanyang Business School Research Paper No. 20–31, SSRN (2020)
Matthews, D.: A History of Auditing: The Changing Audit Process in Britain from the Nineteenth Century to the Present Day. Routledge, London (2006)
Meiryani, B., Susanto, A.: The influence of information technology on the quality of accounting information system. In: Proceedings of the 2nd High Performance Computing and Cluster Technologies Conference, pp. 1–6. ACM, Beijing (2018)
Mutchler, J.F., Hopwood, W., McKeown, J.M.: The influence of contrary information and mitigating factors on audit opinion decisions on bankrupt companies. J. Account. Res. **35**(2), 295–310 (1997)
Nasrah, H., Muda, I.: Kesuma: computer assisted audit tools and techniques adoption: a systematic literature review. Int. J. Soc. Serv. Res. **3**(3), 630–638 (2023)
Omonuk, J.B., Oni, A.A.: Computer assisted audit techniques and audit quality in developing countries: evidence from Nigeria. J. Internet Bank. Commer. **20**(3), 1–17 (2015)
Pedrosa, I., Costa, C.J., Aparicio, M.: Determinants adoption of computer-assisted auditing tools (CAATs). Cogn. Technol. Work **22**(3), 565–583 (2019)
Purnamasari, P., Hartanto, R.: Determinants of effectiveness of computer-assisted audit techniques in the public sector. Probl. Perspect. Manag. **20**(4), 250–263 (2022)
Robson, K., Humphrey, C., Khalifa, R., Jones, J.: Transforming audit technologies: Business risk audit methodologies and the audit field. Acc. Organ. Soc. **32**(4–5), 409–438 (2007)
Singleton, W., Singleton, A., Bologna, J., Lindquist, R.: Fraud Auditing and Forensic Accounting, 3rd edn. Wiley, New York (2006)
Zhang, C.: Intelligent process automation in audit. J. Emerg. Technol. Account. **16**(2), 69–88 (2019)

The Impact of AI-Assisted Translation on Corporate Culture in Multinational Companies: A Multi-theoretical Overview

Abdelkrim Chirig

Abstract Multinational companies operate in multilingual environments where language policy shapes participation, trust, and professional identity. In this context, AI-assisted machine translation increasingly enables routine cross-lingual communication through instant messaging, document circulation, and real-time meeting support, offering an alternative to English-only regimes and costly human mediation. Accordingly, this chapter provides a multi-theoretical overview of how AI translation reshapes corporate culture and organizational communication. Drawing on sociotechnical systems theory, technological determinism, Hofstede's cultural dimensions, media richness theory, linguistic relativity, language ideology, and institutional theory, it explains why the same technology can democratize employee voice and knowledge sharing in some contexts while reproducing hierarchy and mistrust in others. More specifically, the discussion highlights shifts in power relations, including reduced reliance on bilingual gatekeepers, new competencies such as machine translation literacy and machine-oriented writing practices, and changing headquarters–subsidiary knowledge flows. At the same time, it examines risks that may weaken cohesion, including mistranslation, loss of interpersonal nuance, technostress, exposure to privacy and confidentiality risks, and the centralization of control through translation governance and vendor dependence. Ultimately, the chapter concludes that AI translation is a cultural intervention whose organizational effects depend on context-sensitive adoption and trust-building practices.

Keywords AI-assisted translation · Corporate culture · Language ideology · Multinational companies · Multilingual communication

A. Chirig (✉)
Chouaib Doukkali University, El Jadida, Morocco
e-mail: chirig.a@ucd.ac.ma

A. Bouziane and K. Bouziane (eds.), *AI, Management, and Culture: Shaping Organizations in the Digital Age*, Studies in Computational Intelligence 1291,
https://doi.org/10.1007/978-3-032-27882-1_11

1 Introduction

In today's globalized business environment, multinational companies operate in multilingual settings, posing long-standing challenges for organizational communication and culture (Harzing et al. 2011; Tenzer et al. 2014). Historically, firms addressed language barriers by adopting a single corporate lingua franca (often English) or by hiring human translators. To elaborate, the Japanese e-commerce company Rakuten enforced English-language proficiency through a radical *Englishnization* initiative. Although it increased international coordination, it led to productivity deficits and greater employee stress due to the use of a different language (Neeley 2017; Harzing and Feely 2008). The example of Rakuten illustrates how language policy choices affect business culture, power relations, and employee identity.

Many AI-assisted translation technologies now offer an alternative means of bridging language gaps. These tools can instantly translate text or speech between dozens of languages, enabling individuals to communicate in their native languages across linguistic boundaries. The rise of AI translation promises unprecedented inclusivity and efficiency in global collaboration. Employees can read company documents or converse with colleagues in their preferred language with machine assistance, rather than being required to switch to a dominant language such as English. This technological achievement also has the potential to significantly impact the culture of global institutions, including patterns of knowledge sharing, governance and power distribution, the self-perception of professional identity, and the political implications of valuing or depreciating languages (Koskinen 2020; Moorkens 2017).

This chapter, hence, examines how AI-assisted translation is reshaping corporate culture and communication across sectors and regions. We draw here on multiple theoretical frameworks to analyze the issue from different angles: sociotechnical systems theory to consider the interplay of technology and social structures (Bostrom and Heinen 1977), technological determinism to ask whether technology drives cultural change or vice versa (Winner 1980), Hofstede's cultural dimensions to explore cross-cultural differences in communication preferences (Hofstede 2001), media richness theory (Daft and Lengel 1986) to evaluate the quality of machine-mediated communication, linguistic relativity to assess how language influences thought and whether meaning is lost in translation (Lucy 1997), language ideology theory to understand beliefs and power relations tied to language use (Kroskrity 2004), and institutional theory to see how organizational norms and pressures affect the adoption of AI translation (DiMaggio and Powell 1983). Therefore, by adopting these perspectives, the focus shifts to analyzing the impacts of AI translation on power relations, knowledge hierarchies, identity, and language politics within organizations. The chapter also critically evaluates the benefits and drawbacks of AI-assisted translation for international corporate culture, integrating conceptual insights with practical examples. The ultimate goal is to provide a comprehensive understanding of this emerging phenomenon.

2 Theoretical Foundations: Technology, Language, and Organizational Culture

To establish a foundation for the current analysis, we first outline key theoretical frameworks that illuminate the intersection of AI translation technology and organizational culture. And each framework provides concepts for examining how the introduction of AI-assisted translation may affect communication patterns, social relationships, and cultural norms within organizations.

2.1 Sociotechnical Systems and Technological Determinism

Sociotechnical systems theory, on the one hand, views organizations as joint systems of social elements (people, culture, and structures) and technical elements (tools and processes) that must be aligned for optimal outcomes. From this perspective, AI translation is not merely a neutral add-on, but rather part of an integrated human–machine system whose effects depend on how well the technology is embedded in users' workflows, training, and the broader social context.

Technological determinism, on the other hand, views technology as a largely autonomous force that shapes organizations in a unidirectional manner. A strong determinist might assume that advanced machine translation will inevitably standardize corporate communication or erode human language skills (Lee 2022). Early philosophers, such as Heidegger, warned that technology enframes or dominates humanity, regardless of its intent (Heidegger 1977). Contemporary scholars, nonetheless, argue that technology is not necessarily good nor bad, although it is not neutral (Kranzberg 1986; Moorkens et al. 2025); we have to keep in mind that its effects depend on how it is used and integrated. Furthermore, technology really comes to life through how we use it, as well as the important role humans play in shaping it. Adopting this balanced view, we treat AI translation as a powerful capability whose cultural impact is filtered through its sociotechnical context, even as we remain alert to structural tendencies introduced by the technology (for example, faster and more frequent cross-lingual exchanges). Together, sociotechnical and deterministic perspectives encourage us to consider whether shifts in corporate culture come from intentional organizational choices or from the inherent possibilities of AI translation itself.

2.2 *Cultural Dimensions and Media Richness in Communication*

Cross-cultural management research, particularly Hofstede's cultural dimensions, provides a second lens. Specifically, differences in Power Distance, Individualism–Collectivism, Uncertainty Avoidance, and other dimensions shape how employees communicate and how they respond to technologies like AI translation. For example, in high Power Distance contexts, language barriers may further inhibit speaking up (Gao and Kang 2019). In this regard, AI translation could either reduce this anxiety by enabling native-language communication (Gajšt 2025) or, conversely, increase it if employees mistrust how respect and formality are rendered. Likewise, high Uncertainty Avoidance cultures may be especially cautious about the imperfections of AI output, thereby preferring established lingua francas or human translators (Alamri 2025). Meanwhile, Individualist versus Collectivist orientations might influence whether such tools are adopted individually or via group consensus (Ali and Abu-Saad 2024). These dynamics are particularly salient in Arab contexts (including Morocco), where relatively higher power distance and collective or tribal structures shape hierarchical, group-oriented communication (Aldawsari 2024). Consequently, we later examine how such cultural profiles condition the reception of AI translation in Arab companies.

Media Richness Theory (Daft and Lengel 1986) categorizes communication media by their capacity for immediate feedback, multiple cues, personalization, and language variety. Thus, face-to-face interaction is considered *rich*, whereas an email is *lean*. Accordingly, managers are expected to match a medium's richness to the complexity of the communication task. However, AI translation complicates this picture. On the one hand, text-based machine translation can further lean out communication by stripping away tone or cultural nuance. For example, a translated email may miss humor, idioms, or relational subtleties. On the other hand, real-time speech translation (e.g., live captioning or simultaneous interpretation bots) can add richness to traditionally lean channels by providing immediacy and multi-sensory cues (Yun and Jung 2022). Therefore, media richness theory prompts us to ask whether AI-mediated communication is sufficiently rich for complex tasks such as negotiation or performance feedback. Relatedly, do organizations develop compensating practices to address their limits, for example, human follow-ups, confirmation routines, or encouraging *MT-friendly* writing norms with simpler syntax and more precise phrasing (Geiger 2019)? In our analysis, Hofstede's dimensions anchor cultural variability, while media richness theory helps evaluate the quality of communication under AI translation (Sangwa et al. 2025).

2.3 Linguistic Relativity and Language Ideology

Linguistic relativity, associated with the Sapir-Whorf hypothesis, argues that language influences thought and perception (Mei 2024). Although strong linguistic determinism remains debated, evidence suggests language shapes memory, categorization, emotion, and risk perception (Sadiqzade 2025). In corporate contexts, meaning is not fully separable from language. For instance, when AI translates a Japanese manager's message into English, it may retain core content yet subtly alter tone, politeness, or urgency. In particular, honorifics, indirect phrasing, and culturally specific concepts often lack neat equivalents, which can, in turn, introduce distortion. At the same time, AI translation can let employees communicate in their strongest language, thereby lowering cognitive load and enabling more authentic expression. Consequently, it may surface ideas that might be suppressed under an enforced lingua franca (Atarchi et al. 2024).

Language ideology theory focuses on shared organizational beliefs about language. In particular, ideologies such as (English signals competence) or (local languages must be maintained at work) shape willingness to speak and perceived legitimacy. In this sense, language ideologies can be defined or considered as shared sets of beliefs about languages in society, and they can create participation barriers. For example, an English-only perfectionism norm can silence non-native speakers, whereas, by contrast, normalizing code-switching can support inclusion (Barner-Rasmussen et al. 2024). In this context, AI translation intersects with these beliefs: if machine output is viewed as inferior, employees may resist it; however, if multilingualism and technological pragmatism are valued, AI tools may be treated as equalizers. Nevertheless, tensions can arise when management frames AI as *good enough* and cost-efficient, while translators or local managers see it as a threat to quality or to the status of certain languages.

Language ideologies also link to power and identity. Prestige languages, often headquarters or global business languages such as English or French, can shape career opportunities and perceived professionalism (Roncevic 2024). In this sense, AI translation may broaden participation for monolingual staff; however, machine-mediated contributions can still be discounted as less credible (Tonkin 2020). Moreover, institutional policies may reinforce or challenge these hierarchies: for instance, some firms require multilingual communication via AI, while others treat machine translations as unofficial unless human-vetted, thereby preserving the primacy of the original language (Lønsmann and Kraft 2018). Because workplace policies often privilege dominant languages even under the banner of inclusion, translation practices can reproduce linguistic inequality, including when translations are structurally positioned as secondary (Angermeyer 2022). In North Africa, French in Morocco retains neocolonial and elite connotations, while Arabic is tied to national identity (Chakrani and Huang 2014). Consequently, AI translation can become an ideological battleground, either empowering Arabic in corporate life or indirectly stabilizing global-language dominance by further extending its reach.

2.4 *Institutional Theory and the Diffusion of AI Translation*

Institutional theory suggests that organizations adopt new practices not only for efficiency but also to gain legitimacy by conforming to broader norms and expectations (DiMaggio and Powell 1983). In the case of AI translation, institutional pressures include mimetic forces (imitating competitors and industry leaders), normative forces (emerging professional norms favoring multilingual accessibility and technological modernity), and coercive forces (regulatory or client requirements for multilingual documentation or state policies promoting AI adoption). As a result, implementing AI translation can signal innovation or commitment to diversity and inclusion, rather than merely solving operational problems. Over time, AI translation may become institutionalized through formal roles, training programs for machine translation literacy (Bindels et al. 2024), and standardized procedures. Nevertheless, institutional theory also warns about decoupling: organizations may symbolically adopt the tool but fail to integrate it meaningfully, thereby keeping old language habits despite new AI infrastructure.

Institutional perspectives highlight shifting power relations as well. Historically, multilingual gatekeepers, including professional translators and bilingual employees, controlled cross-language communication, holding *hidden power* to filter and reframe messages (Massey 2021). In our context, AI translation can displace these actors if managers bypass them using automated systems, thereby redistributing power toward those who control the technology (IT departments or external vendors). In that case, the authority to shape meaning partly migrates to algorithms and the corporations that own them, which raises concerns about concentrating technology and power in a small number of multinational companies (Moorkens et al. 2025). Further institutionalization may appear in policies requiring that all internal communications be available in multiple languages via AI translation. Such policies advance multilingual inclusion and ensure legal or compliance needs are met; notwithstanding, they also introduce oversight questions: Who checks the quality and fixes errors? Who handles culturally sensitive content that the machine might misinterpret? Institutional theory thus draws attention to the formal rules, roles, and external dependencies that accompany the diffusion of AI translation.

3 AI-Assisted Translation in Multilingual Organizations: A Cross-Sector Overview

Before examining cultural impacts, it helps to note how AI-assisted translation is used across industries and regions. In practice, multilingual organizations, including multinational firms, international NGOs, and distributed tech companies, increasingly integrate AI tools into routine communication, ranging from documents and emails to real-time meeting support. Modern systems, often based on neural networks and large language models, provide fast translation across many languages at much higher quality than earlier machine translation, while also operating at very low marginal cost.

In manufacturing, construction, and other high-risk workplaces, studies link limited proficiency in the dominant workplace language to safety communication breakdowns and elevated risk, especially for migrant or non-dominant language workers. In addition, they document mitigation strategies, such as translating safety materials and adapting communication for multilingual crews (Lindhout et al. 2019). This aligns with occupational safety guidance indicating that training should use language and vocabulary employees can understand, with designs that support comprehension (Occupational Safety and Health Administration 2010; Occupational Safety and Health Administration 2015; Occupational Safety and Health Administration 2021). In knowledge-intensive and globally distributed organizations, language barriers can undermine trust and disrupt team processes (Tenzer et al. 2014), while stronger internal communication is associated with higher employee engagement (Karanges et al. 2015). Taken together, this supports the expectation that AI translation, used to localize safety documentation, onboarding, and internal communication, can improve comprehension and inclusion, though outcomes should be evaluated in context.

Healthcare is also adopting AI translation for critical content. For example, a European healthcare organization translated compliance training and policy documents into five languages, improving understanding and reducing training costs through centralized automation (Genovese et al. 2024). More broadly, cost reduction is a recurring driver across sectors: AI can produce first-pass translations while human experts focus on fine-tuning and high-stakes content, thereby reducing reliance on extensive human translation capacity and limiting the burden on bilingual staff. In addition, customer service also uses real-time translation so monolingual agents can support foreign-language customers, potentially reducing the premium historically attached to bilingual workers. In Morocco, call centers are beginning to use AI so an Arabic-speaking agent can assist a French-speaking customer in real time, and vice versa. Consequently, this could broaden recruitment pools, but it may also reshape status and pay structures that have traditionally rewarded multilingual proficiency (Frey and Llanos-Paredes 2025).

Morocco and the wider Arab world are a revealing setting because workplaces are deeply multilingual, with Darija and MSA alongside French, English, and sometimes Amazigh, which can create bottlenecks and high-status language hierarchies; that is, AI translation can ease these through real-time exchange and more accessible training. Additionally, evidence from North African factories suggests that delivering training in workers' strongest languages is associated with better outcomes, about 8% turnover and 5% incident rates, compared with 18 and 12% under English-only programs (Matsh 2025). Then again, adoption often begins with uninformed use of free tools, increasing confidentiality and error risks (Bindels et al. 2024). For this reason, scholarship emphasizes MT literacy and critical awareness, and it shows how professional translators may resist or negotiate MT use in practice, pointing to the need for informed, skilled engagement (Cadwell et al. 2018). Consequently, many firms, therefore, combine secure MT, human review, and staff training to support quality and trust (Zavala-Rojas et al. 2024).

Industry guidance is also emerging. For instance, post-editing standards (such as ISO 18587:2017) outline how humans should correct MT output (Massey 2021). As a result, in-house translators often shift toward post-editing or advisory roles, which can affect professional identity and status by moving from authoring to editing (Olohan 2021). At the same time, teams also experiment with new workflows, for example, drafting in their strongest language, translating into a shared language for group review, or using live translated captions in virtual meetings on platforms such as Zoom and Microsoft Teams. In practice, early adopters report benefits but also limitations, since jargon and strong accents can reduce accuracy, and some cultural nuance can be lost, requiring clarification in discussion (Briva-Iglesias et al. 2023).

4 Shifting Power Relations and Knowledge Hierarchies

Language in organizations is closely tied to power and control over knowledge. In many settings, fluency in the dominant corporate language, often English, gives direct access to top management and strategic information, while less fluent employees depend on intermediaries, which creates hierarchies and bottlenecks (Boussebaa et al. 2014). In this context, AI-assisted translation can reshape these dynamics by changing who is empowered and how knowledge moves (Zhang et al. 2022).

A key shift is the weakening of traditional language gatekeepers. In many MNCs, expatriate or bilingual local managers have acted as communication hubs, and their ability to filter and interpret information has been a source of power. With advanced AI translation, HQ can communicate directly with staff in their own languages, and employees can submit ideas in Arabic or Spanish that HQ reads in English or French. This matters because language barriers often block idea-sharing, and subsidiary knowledge can be lost in translation before reaching decision-makers (Worakittikul et al. 2025). Moreover, studies show that language barriers can suppress employee voices, limiting participation in innovation to fluent employees (Guillouet et al. 2021). By lowering barriers, AI translation can enable bottom-up idea flow. For example, factory-floor employees in Latin America used an internal MT tool to submit Spanish suggestions directly to a global engineering team, resulting in hundreds of proposals and measurable efficiency gains, consistent with evidence that better communication and fewer language barriers support employee-driven innovation and empowerment (Koech et al. 2024). In this way, AI translation can reduce reliance on language-skilled intermediaries and soften hierarchies.

Yet power does not disappear; it is reconfigured. For example, bilingual managers and professional translators may feel their status threatened when AI translates in seconds; consequently, some reposition themselves as post-editors or cultural consultants, while others resist adoption by stressing errors or discouraging use. In this context, the hidden power of translators (Vanmassenhove 2020) may shift toward less visible actors, particularly algorithm developers and those who control training data. Moreover, if languages or dialects are underrepresented, their speakers remain marginalized (Zaki and Ahmed 2024). Standard Arabic is relatively well supported, but many Arabic dialects still yield weaker results unless systems are trained on them (Seyidov 2024), which can distort voices, force switching to prestigious languages like French or English, and reproduce digital linguistic inequality.

Accuracy and control create further power issues. For example, when management relies heavily on AI to translate local messages, nuance can be lost; therefore, organizations often require clear, simple writing for machine translation and double-checking of critical texts (Ningrum et al. 2024). Over time, employees develop machine-mediated communication competence, learning to phrase messages so that AI handles them reliably; as a result, this becomes a valued skill (Rajagopal and Lakshmi 2025).

AI translation also reshapes HQ–subsidiary knowledge hierarchies. In many cases, HQ–subsidiary knowledge transfer is hindered by language differences (Tenzer and Schuster 2017). For example, a field experiment in Myanmar found that English training for local middle managers increased interaction with foreign senior managers and involvement in decision-making (Guillouet et al. 2021), which suggests that reducing language barriers increases local agency. In a similar way, AI translation can mimic this by reducing friction immediately; however, the same study warns that lower barriers can also allow HQ to push information downward more quickly, thereby reinforcing one-way control if listening does not increase.

Trust is central. In many settings, language gaps can undermine interpersonal trust in teams. In response, AI translation can improve mutual understanding and strengthen trust; however, the latter, in machine-mediated contexts, is fragile, as awkward or misleading translations can fuel rumors and require clarification (Robertson and Díaz 2022). To reduce this risk, some organizations show originals and translations side by side and remain transparent about what was post-edited by bilingual staff (Vössing et al. 2022).

Meetings illustrate these shifts. In single-language multilingual meetings, native speakers tend to dominate while others struggle to contribute and may retreat into side talk in their own language (Śliwa and Johansson 2014). In this context, real-time AI interpretation can enable participants to speak in their strongest language and receive translations, thereby reducing the native-speaker advantage and allowing expertise, rather than fluency, to drive influence (Santorelli and Catullo 2023). Early outcomes are mixed, with more inclusivity but slower pacing and some loss of humor or subtlety (Zhang et al. 2022). Nevertheless, this may normalize multilingual meetings and challenge the hegemony of English in international business.

Finally, power increasingly concentrates on the control of AI systems themselves. In particular, external providers shape translation quality, style, and terminology (Mohamed et al. 2024). In response, some firms implement custom MT systems and company glossaries to standardize key terms (Sandrini 2025). As a result, teams governing these systems gain influence by deciding which languages to support, how data is managed, and what quality thresholds apply (OECD Digital Economy Papers 2023). In this respect, governance can empower when regions have a voice, or it can become centralized and biased when HQ alone decides (Angermeyer 2022).

For example, a U.S. firm might optimize English–Spanish and English–Chinese but underinvest in English–Arabic or French–Arabic, leading to lower-quality translations in Arab countries and implying those languages and contributions are less valued (Li 2025). Technical choices thus become political choices in organizational culture.

5 Identity, Inclusion, and Language Politics in the Workplace

The languages we speak are central to our personal and social identity. In multinational workplaces, language groups often align with cultural groups, and employees frequently cluster socially with those who share their native language. As a result, a common language provides comfort and trust, reinforcing what researchers call linguistic clustering in organizations (Kalra and Danis 2024). At the same time, the choice of workplace language is never neutral: it involves language politics, namely struggles over whose language is given official status, whose communication styles are deemed professional, and which languages are marginalized (Wilmot et al. 2023). In this context, AI-assisted translation enters this socio-cultural terrain as a potentially disruptive force. Consequently, it raises critical questions: Will it allow people to bring their complete linguistic identity to work (speaking in their native language rather than in a second language), thereby enhancing inclusion and authenticity (Csata and Marácz 2021)? Or will it create a superficial homogenization, in which everyone's words are filtered into a single corporate monolingualism, potentially eroding the rich tapestry of linguistic identity (Maitland et al. 2022)? Also, how will existing language politics, especially in regions such as the Arab world with colonial language legacies, play out when translation is readily available?

5.1 Identity and Speaking One's Own Language

For many employees, using their native language at work brings pride and comfort, whereas working in a second or third language can make them feel that their personality or voice is diminished. As a result, they may avoid jokes, stories, or emotional expression because these do not transfer well (Jeong and Lindemann 2024). This can create a split identity between a professional persona, such as an English-speaking international self, and a truer self in one's native language. In our context here, AI translation can narrow that gap and affirm linguistic identity by providing employees with space to use their first language (Anik et al. 2025). This can increase engagement and job satisfaction when employees feel their identity is respected (Courtney and Phelan 2019), thereby supporting broader inclusion goals.

At the same time, identity expression may change when communication passes through an impersonal interface. For example, employees sometimes feel that machine-translated text does not reflect their voice because the tone can become uniform or generic (Long 2022). This reflects a known limitation: many MT systems prioritize meaning over the preservation of dialect, register, or personal style (Aldawsari 2024; Miller et al. 2018). As a result, to achieve predictable outcomes, employees may simplify and standardize their language, thereby self-censoring culturally specific or creative expression (Kim and Yi 2025). Because style and code choices are part of performing identity, losing them can subtly reduce how connected or maybe heard people feel (Xiao and Hancock 2025).

AI translation also challenges workplace language ideologies tied to corporate identity. In many multinationals, identity has historically been built around a primary corporate language, for example, by framing English as a symbol of global outlook. For instance, some Japanese firms explicitly linked adopting English to a forward-looking orientation, and Rakuten's shift is often discussed in these terms (Kubota 2013). If AI makes everyday multilingualism feasible, corporate identity may come to regard multilingualism itself as a core value (Ishikawa 2022). This can be positive; however, it may also create uncertainty about a linguistic center and about whether teams become language silos connected primarily through tools. One balancing approach is to maintain a lingua franca for specific domains, such as leadership addresses or official records, as the primary reference, while encouraging multilingual communication in day-to-day communication (Leal 2021).

Employee ideologies matter too. In contexts such as North Africa, workers may perceive a language ceiling in which limited proficiency in French or English impedes promotion, thereby producing resentment and exclusion. In this regard, the Matsh (Matsh 2025) report on North African factories found that monolingual workers felt English-only training "sends negative messages about whose knowledge counts" (para. 14), which undermines trust and collaboration, and the phrase "whose knowledge counts" (Matsh 2025), para. 14 captures how language signals whose voices are valued. Moreover, in Matsh's data, factories that adopted multilingual training and communication reported higher retention and performance, thus suggesting that linguistic inclusion strengthens commitment and effectiveness (Matsh 2025). Because language is a marker of belonging (Burley 2023), AI translation in such settings can reframe status: workers' knowledge counts because it can be conveyed in their language and still understood by management.

However, some worry that effortless translation could reduce incentives to build cross-cultural understanding. For example, if everyone relies on a tool, groups may become siloed and connect only through machines (Carvalho et al. 2023). In global teams, adopting a shared lingua franca has historically forced mutual adaptation and sometimes produced hybrid team cultures and camaraderie (Szczepaniak-Kozak et al. 2022). If AI replaces that effort, the motivation to learn colleagues' languages, often seen as a sign of respect and a relationship-deepening gesture, may decline (Muñoz-Basols et al. 2023). In this vein, this tension between technological convenience and human connection remains open.

Generational differences can intensify these dynamics. To illustrate this idea, younger digital natives in places such as Morocco may be more willing to trust AI and to assert local languages, whereas older executives may prefer French or English and distrust machine quality, which can produce ideological clashes (Zakhir and O'Brien, 2019). Furthermore, early experiences can shape the narrative: reliable performance can normalize adoption, but high-profile errors can reinforce resistance. For this reason, some firms conduct controlled pilots to avoid early, high-stakes mistakes and to build confidence gradually (Ennaji 2021).

Overall, AI-assisted translation can strengthen identity expression and inclusion by enabling native-language communication and validating those languages within corporate culture (Anik et al. 2025). In addition, it can weaken the link between esteem and foreign-language proficiency, potentially challenging historically privileged languages, while real-world multilingual programs have shown gains, including improved comprehension and lower turnover (Matsh 2025). However, organizations must manage new complexities, including preserving personal voice (Aldawsari 2024; Long 2022; Miller et al. 2018), preventing siloed communication (Carvalho et al. 2023), and aligning differing language ideologies through careful rollout (Ennaji 2021). With that balance, multilingualism can shift from being treated as a problem to be solved to a core element of corporate identity.

Having examined power, knowledge, identity, and politics, the next step is to synthesize these insights by weighing the benefits and drawbacks of AI-assisted translation for international corporate culture and communication. In doing so, it becomes possible to clarify where value lies and what must be managed for successful sociotechnical integration.

6 Benefits and Drawbacks of AI-Assisted Translation for Corporate Culture

As we have seen, AI-assisted translation technology can profoundly influence the cultural fabric of international companies. Its adoption brings a mix of promising benefits and notable drawbacks. In this section, we critically evaluate these pros and cons, drawing on both theoretical reasoning and documented cases. The aim is to provide a clear-eyed assessment to guide managers and stakeholders in leveraging AI translation effectively while remaining mindful of its limitations.

6.1 Key Benefits

Enhanced Inclusivity and Equity. AI translation allows every employee to access essential information in their own language, thereby fostering greater inclusion and equity. No longer are important announcements or documents limited to those fluent in the corporate lingua franca; instead, everyone can read and respond in the language they are most comfortable with. This leveling of the field enables more equitable participation in meetings and decision-making (Güven et al. 2025). In turn, workers who previously stayed silent due to language barriers can now have a voice through translation. Studies confirm that overcoming language barriers is essential for fostering an innovative, empowering culture in multinational firms (Worakittikul et al. 2025). In short, AI translation helps democratize communication by tapping the full diversity of insights within a company.

Improved Knowledge Sharing and Collaboration. Removing language barriers improves knowledge flow across the organization. As a result, expertise and best practices can circulate more freely between units once translation is no longer a bottleneck. Accordingly, AI translation enables teams in different countries to share reports and build on one another's ideas without delay or misunderstanding, thereby preventing knowledge from being trapped in language silos (López-Arroyo et al. 2024). Empirical evidence, in turn, shows that reducing language frictions led to more frequent interactions between foreign and local managers and greater involvement of local staff in decision-making (Guillouet et al. 2021). On the contrary, when language barriers persist, they often lead to a polarization of perspectives and the hoarding of knowledge within language groups (Worakittikul et al. 2025). In this regard, AI translation acts as a bridge, encouraging cross-pollination of ideas and collaborative problem-solving across linguistic divides. For example, multilingual project teams using these tools have reported reaching consensus more quickly because misunderstandings were minimized, and each member could contribute technical details precisely in their own language. There is also a potential boost to creativity: language diversity, supported by translation, can spur new ways of thinking and innovation (Almeida 2023).

Preservation of Linguistic and Cultural Identity. Instead of imposing a one-size-fits-all language policy, companies using AI translation can adopt a truly multilingual ethos that respects and preserves employees' cultural identities. This has qualitative value, improving morale and organizational commitment. Employees feel seen and respected when they do not have to hide their native language or pretend to be someone they are not linguistically. Particularly in regions emerging from colonial histories, such as parts of the Arab world, using one's mother tongue at work is deeply empowering and symbolically significant (Boussebaa et al. 2014). It signals that heritage is not a barrier to advancement. From a knowledge perspective, certain local concepts or insights may be fully expressible only in the local language; translation allows those culturally embedded ideas to surface, thereby enriching the company's knowledge base. This aligns with linguistic relativity, which holds that language influences thought (Lucy 1997). Therefore, embracing multiple languages captures multiple ways of understanding business challenges.

Efficiency and Speed of Communication. AI translation dramatically accelerates processes that were previously slow. In the past, getting a document translated might have taken days of turnaround via human translators or bilingual colleagues. It can now often be completed in seconds or minutes. As a result, multinational operations can proceed more quickly. For example, product manuals can be localized promptly for different markets, global HR can roll out a new policy simultaneously worldwide in all needed languages, and urgent technical fixes can be shared in real time. The result is quicker decision-making because comprehension is instant. In a crisis, for instance, if a safety issue is identified in one country, AI translation enables immediate alerts and instructions to be issued in all relevant languages, potentially preventing accidents. Properly deployed machine translation in crisis scenarios "can save lives" [17, para. 1]. Even in routine situations, the efficiency gains are clear, and companies have reported time savings in translating standard materials (Shahmerdanova 2025). This efficiency can indirectly improve culture by reducing frustration. Employees are not left waiting on a translation to understand what HQ is asking, and managers are not frustrated that field offices did not get the memo. When people communicate promptly, it strengthens alignment and avoids the mistrust that delays and silence can sow.

6.2 Key Drawbacks

Despite these benefits, some significant drawbacks and challenges can accompany AI-assisted translation and negatively impact corporate culture if not addressed.

Risk of Miscommunication and Errors. AI translation, while powerful, is not infallible. As a result, errors or subtle mistranslations can undermine communication and trust. Especially worrisome are mistakes in critical information, since a mistranslated safety warning or contract clause can have serious consequences. In addition, cultural nuances are often lost or mistranslated by machines: what one party intended as a polite suggestion might be interpreted as a rude directive after translation, potentially causing offense; or a joke might be translated literally as something nonsensical or inappropriate (Al-Kaabi et al. 2024). If the error rate is high or unpredictable, it can foster a culture of doubt, and people may not fully trust what they read or hear via AI, which can lead them to hesitate to act on it (Duan et al. 2024). To mitigate this, companies often keep a human-in-the-loop for high-stakes content and encourage a feedback loop in which employees can flag anomalous translations (Kapoor 2025). Nevertheless, until AI translation achieves near-human understanding of context and tone, this drawback remains a trade-off between speed and absolute accuracy. From a cultural perspective, organizations must be prepared to manage the fallout of mistakes or risk a loss of trust in the tool.

Overreliance and Skill Atrophy. Another critical drawback is the potential overreliance on AI translation, which can erode human bilingual skills and foster complacency in intercultural communication. If employees come to depend entirely on the tool, they may stop making any effort to learn colleagues' languages or improve their own second-language proficiency. Bilingual or multilingual employees often serve as cultural bridges, not just translating language but also explaining context and mediating understanding between groups (Gonçalves and Schluter 2017). If the mindset becomes that *the AI will translate*, human mediation and the learning that comes with it may diminish. This is risky because machine translation, no matter how advanced, cannot fully replace humans in all scenarios, particularly in nuanced negotiations, sensitive conflict resolution, or creative discussions where cultural context matters (Liddicoat 2016). If the technology fails or is unavailable due to an outage, a policy change, or a vendor issue, communication barriers could re-emerge without human buffers to bridge them. In short, a company could become linguistically *lazy*, which might hinder deeper integration of its multicultural teams. Some organizations mitigate this by adopting a hybrid approach: they use AI while maintaining traditional language-learning and cultural-exchange programs, so that human skills and understanding continue to develop alongside the technology (Shahmerdanova 2025).

Impersonal Communication and Reduced Relationship Depth. Communicating via a machine intermediary can, at least initially, feel impersonal or transactional. There is concern that, if colleagues primarily interact through translated text or synthesized speech, some of the warmth and spontaneity of human interaction may be lost (Harris-Watson et al. 2023). People often bond over small language exchanges, such as teaching each other phrases or laughing over mistakes. With near-perfect translation, those moments of shared human vulnerability might vanish. Additionally, tone of voice and personality can be flattened by the uniformity of machine output or plain text (Ayub 2025). This could make communications more sterile and strictly task-focused, potentially undermining the informal social interactions that build team camaraderie (Arakelian and Rudolfsson 2023). In this respect, we should consider media richness: a translated text or AI-generated voice generally has lower social presence than a direct human voice full of emotion (Bergin 2013). Consequently, employees may be less likely to engage in cross-language casual banter if it does not have the same effect. To illustrate, making a friendly joke to an international colleague is risky if you are not sure the translator will capture the tone, so you might skip it, thereby sticking only to business matters (Es-sousy et al. 2024). Over time, these small omissions could add up to a more formal, distant communication culture across language groups. In the worst case, people may come to see one another as machine-mediated *avatars* rather than full human colleagues, which could reduce trust and empathy. This scenario is speculative, since human nature tends to find ways to connect, but it is a noted worry, especially in relationship-centric or high-context cultures (Merchant and Adnan 2025). If translation errors or monotony strip away relational cues, they can hamper the development of a team family feeling across cultures (Abu-Rayyash 2023).

Maintaining Power Imbalances Through Technology. We noted earlier that AI translation can shift power, but it can also reinforce existing power imbalances under a new guise; that is, if AI deployment is controlled by a central authority, such as when HQ decides which languages receive robust translation support, or if the system performs significantly better for some languages than others, it may perpetuate a subtle linguistic hierarchy. Suppose an MNC provides excellent machine translation for English–Spanish and English–French, the languages of top management, but mediocre support for English–Arabic. In essence, biases in technology, often data-driven and favoring well-resourced languages, can mirror those of social systems (Ciston 2019). Institutional theory reminds us that firms may adopt the façade of inclusivity, translating everything, while still maintaining centralized control (Li & Freeborn 2025). Besides, management may gain greater surveillance capabilities, as they can now read everything employees say, in any language. If misused, AI translation could become a tool for centralized control or surveillance, thereby worsening power asymmetries rather than alleviating them (Ishkhanyan 2025).

Security and Confidentiality Concerns. From a cultural perception, trust in the system also requires confidence that sensitive information will remain secure. Many AI translation services are cloud-based; therefore, if employees use them carelessly, they may inadvertently expose confidential plans or personal data. A data leak or privacy breach can severely damage organizational culture by eroding employees' confidence in management's judgment and creating an atmosphere of fear around communication (Posey et al. 2010). In other words, if HR shares employee grievances or performance reviews via an unsecured, free service and a leak occurs, employees will feel their privacy was violated. Similarly, if R&D texts are translated using a public engine and components are cached externally, this can create paranoia: employees may become reluctant to document sensitive information, even internally, for fear that it could pass through a translation tool and be exposed. This risk is not inherent to AI translation itself, but rather to its improper use without safeguards. It is a drawback because it requires proactive measures, such as using secure, enterprise-approved MT solutions and clear policies; otherwise, a breach could undo cultural gains by introducing mistrust.

Technostress and Resistance. Introducing any new technology can cause technostress, namely the stress of learning and trusting new systems (Tarafdar et al. 2019). As a consequence, some employees may find using AI translation cumbersome or intimidating, especially if they are not very tech-savvy. If the interface is not user-friendly or if it adds steps to their routine, they may become frustrated or embarrassed. By way of illustration, an older employee struggling to activate the meeting translator might feel exposed, which can lead to disengagement. There can also be outright resistance from those who feel the translation quality is not good enough or who see it as an unnecessary gimmick. If management pushes the tool too aggressively without listening to feedback, it could hurt morale or create a cultural rift between tech enthusiasts and traditionalists. Another important point is that professional translators or linguists within the company may feel their roles threatened, as discussed earlier, and could either passively resist, for example, by not cooperating with training the AI, or lose motivation. Changes in job roles, such as asking engineers and subject experts to also review translations, could be seen as extra work, thereby breeding resentment. All this highlights that the cultural adoption of AI is not automatic; mishandling the human side can create a toxic environment around the tool, with people mocking it, avoiding it, or using it incorrectly, causing problems. Overcoming this requires good change management: involving employees in pilot testing, providing training, such as MT literacy sessions to build user confidence (Bindels et al. 2024), and creating an environment where feedback on the tool is welcomed and acted upon.

To get the best of both worlds, companies can pair AI translation with occasional language-exchange sessions to deepen mutual understanding. They can also maintain a bilingual glossary, curated by translators or subject experts and integrated into the AI, to keep terminology consistent and culturally appropriate, thereby improving accuracy and trust. Furthermore, they should monitor cultural impact through surveys or focus groups to ensure employees feel included and empowered. When managed this way, the benefits for inclusivity, knowledge sharing, and speed can outweigh the drawbacks. But if risks are ignored, trust and effectiveness may erode, and the tool may be abandoned or lead to quiet dysfunction.

7 Conclusion

AI-assisted translation is rapidly shifting from a peripheral support function to a central communicative infrastructure in multinational companies. As this chapter has shown, its cultural consequences extend well beyond efficiency. By enabling routine cross-lingual messaging, document circulation, and real-time meeting support, AI translation can widen participation, reduce reliance on bilingual intermediaries, and accelerate knowledge transfer between headquarters and subsidiaries. In many settings, this helps democratize voice: expertise can matter more than fluency, and ideas that would previously remain trapped in local languages can enter wider organizational circulation. At the same time, the technology may simplify communication by toning down formality, blending politeness rules, and reducing interpersonal nuances, which could erode trust if misunderstandings build up or if employees perceive their voice as bland and mechanical.

Across the theoretical lenses mobilized in this chapter, a consistent insight emerges: AI translation is not culturally neutral, and its effects are not predetermined. Sociotechnical systems theory clarifies that outcomes depend on how the tool is embedded in workflows, training, and accountability. Technological determinism reminds us that AI introduces structural tendencies, such as faster and more frequent cross-lingual exchanges, yet organizational choices still mediate whether these tendencies empower or control. Cultural dimensions and media richness help explain why the same translation practice may be welcomed in one context and resisted in another, particularly when power distance, uncertainty avoidance, and high-context interaction shape expectations of formality and relational depth. Linguistic relativity and language ideology highlight that translation reshapes meaning, stance, and perceived legitimacy, and therefore can either challenge or reproduce linguistic hierarchies. Institutional theory further explains that adoption may be driven as much by legitimacy and imitation as by functionality, thereby raising the risk of symbolic multilingualism, in which translation exists on paper but does not translate into genuine voice and inclusion.

These shifting aspects are particularly salient in postcolonial multilingual settings such as Morocco, where Arabic varieties, French, English, and Amazigh carry unequal symbolic and career value. In such environments, AI translation can become an ideological battleground: it can strengthen inclusion by validating employees' strongest languages, yet it can also stabilize prestige-language dominance if quality and support remain uneven across languages and varieties. Ultimately, AI-assisted translation can help multinationals move from monolingual norms toward a more inclusive multilingual corporate culture, but only when it is treated as a cultural intervention rather than a purely technical tool. The central challenge for organizations is to capture gains in access, speed, and knowledge sharing while protecting trust, voice, nuance, and equity, so that multilingual communication becomes a durable source of cohesion rather than a new form of invisible stratification.

8 Declarations

The author received no external funding and declares no competing interests. Ethics approval and consent are not applicable (theoretical, literature-based chapter; no human/animal participants). AI-assisted tools were used for language editing/paraphrasing; the author remains fully responsible for the content and references.

References

Abu-Rayyash, H.: AI meets comedy: viewers' reactions to GPT-4-generated humor translation. Ampersand. **10**, 100162 (2023). https://doi.org/10.1016/j.amper.2023.100162

Alamri, J.: Antecedents of generative artificial intelligence technology adoption: extended innovation of diffusion model with cultural dimensions and risks perceptions. J. Ecohumanism. (2025) https://doi.org/10.62754/joe.v4i1.5992

Aldawsari, H.A.H.: Evaluating translation tools: Google translate, Bing translator, and Bing AI on Arabic colloquialisms. Arab World Engl. J. Spec. Issue ChatGPT, 237–251 (2024) https://doi.org/10.24093/awej/ChatGPT.16

Ali, A., Abu-Saad, I.: Quality of work life, cultural values, and Islamic work ethics: the case of the Arab education system in Israel. Int. J. Cross Cult. Manag. **24**, 385–410 (2024). https://doi.org/10.1177/14705958241256729

Al-Kaabi, M.H., AlQbailat, N., Badah, A., Ismail, I.A., Hicham, K.B.: Examining the cultural connotations in human and machine translations: a corpus study of Naguib Mahfouz's Zuqāq al-Midaqq. J. Lang. Teach. Res. **15**(3), 234–249 (2024) https://doi.org/10.17507/jltr.1503.03

Almeida, T.: Language diversity in the workplace can spur creativity and innovation. LSE Bus. Rev.. Available https://blogs.lse.ac.uk/businessreview/2023/02/20/language-diversity-in-the-workplace-can-spur-creativity-and-innovation/ (2023). Last accessed 01 Jan 2026

Angermeyer, P.S.: Translation as discrimination: sociolinguistics and inequality in multilingual institutional contexts. Lang. Soc. **51**(5), 789–812 (2022). https://doi.org/10.1017/S0047404522000422

Anik, M.A., Rahman, A., Wasi, A.T., Ahsan, M.M.: Preserving cultural identity with context-aware translation through multi-agent AI systems. arXiv preprint. (2025) https://doi.org/10.48550/arXiv.2503.04827

Arakelian, E., Rudolfsson, G.: Sharing the same reality: healthy relations between colleagues at work—a meta-synthesis. SAGE Open Nurs. **9**, 1–15 (2023). https://doi.org/10.1177/23779608231207239

Atarchi, K., Elamari, M.M.: The role of artificial intelligence translation tools in academic translation: faculties of pure sciences as a case study. Int. J. Transl. Interpret. Stud. (2024) https://doi.org/10.32996/ijtis.2024.4.3.2

Ayub, A.: Exploring human-AI communication in the digital era: a need to develop new theories for human-AI communication. Int. Soc. Sci. Educ. J. **2**(2), 1–18 (2025) https://doi.org/10.61424/issej.v2i2.181

Barner-Rasmussen, W., Gaibrois, C., Wilmot, N.V.: Perfection, hybridity or shutting up? A cross-country study of how language ideologies shape participation in international business. Int. Bus. Rev. **33**(1), 102189 (2024). https://doi.org/10.1016/j.ibusrev.2023.102189

Bergin, R.: Media Richness Theory (Lecture Script). Center for Homeland Defense and Security, Naval Postgraduate School. Available https://www.chds.us/coursefiles/IS4010/lectures/tech_media_richness_long/story_content/external_files/Media%20Richness%20Theory%20Script.pdf (2013). Last accessed 01 Jan 2026

Bindels, J., Dorst, A.G., Pluymaekers, M.Machine translation in the workplace: deciding on the whether, why and how. Glob. Adv. Bus. Commun. 11(1), 4. Available https://commons.emich.edu/gabc/vol11/iss1/4/ (2024). Last accessed 01 Jan 2026

Bostrom, R.P., Heinen, J.S.: MIS problems and failures: a sociotechnical perspective, part II: the application of sociotechnical theory. MIS Q. **1**(4), 11–28 (1977). https://doi.org/10.2307/249019

Boussebaa, M., Sinha, S., Gabriel, Y.: Englishization in offshore call centers: a postcolonial perspective. J. Int. Bus. Stud. **45**(9), 1152–1169 (2014). https://doi.org/10.1057/jibs.2014.25

Briva-Iglesias, V., O'Brien, S., Cowan, B.R.: The impact of traditional and interactive post-editing on machine translation user experience, quality, and productivity. Translation, Cognition and Behaviour. **6**(1), 60–86 (2023). https://doi.org/10.1075/tcb.00077.bri

Burley, R.: Language is an indicator of belonging. Bonn Search Partners. Available https://bonnsearchpartners.com/language-is-an-indicator-of-belonging/ (2023). Last accessed 01 Jan 2026

Cadwell, P., O'Brien, S., Teixeira, C.S.: Resistance and accommodation: factors for the (non-) adoption of machine translation among professional translators. Perspect. Stud. Transl. Theory Pract. **26**(3), 301–321 (2018). https://doi.org/10.1080/0907676X.2017.1337210

Carvalho, I., Ramires, A., Iglesias, M.: Attitudes towards machine translation and languages among travelers. Inf. Technol. Tour. **25**(2), 175–204 (2023). https://doi.org/10.1007/s40558-023-00253-0

Chakrani, B., Huang, J.L.: The work of ideology: examining class, language use, and attitudes among Moroccan university students. Int. J. Biling. Educ. Biling. **17**(1), 1–14 (2014). https://doi.org/10.1080/13670050.2012.718319

Ciston, M.: Intersectional AI is essential. J. Sci. Technol. Arts. **11**(2), 3–8 (2019). https://doi.org/10.7559/citarj.v11i2.665

Courtney, J., Phelan, M.: Translators' experiences of occupational stress and job satisfaction. Int. J. Transl. Interpret. Res. **11**(2), 1–22 (2019) https://doi.org/10.12807/ti.111201.2019.a06

Csata, Z., Marácz, L.: Social inclusion and multilingualism: linguistic justice and language policy. Soc. Incl. **9**(1), 245–258 (2021) https://doi.org/10.17645/SI.V9I1.3941

Daft, R.L., Lengel, R.H.: Organizational information requirements, media richness and structural design. Manag. Sci. **32**(5), 554–571 (1986). https://doi.org/10.1287/mnsc.32.5.554

DiMaggio, P.J., Powell, W.W.: The iron cage revisited: institutional isomorphism and collective rationality in organizational fields. Am. Sociol. Rev. **48**(2), 147–160 (1983). https://doi.org/10.2307/2095101

Duan, W., Zhou, S., Scalia, M.J., Yin, X., Weng, N., Freeman, G., Gorman, J.C.: Understanding the evolvement of trust over time within human-AI teams. Proc. ACM Hum.-Comput. Interact. **8**(CSCW1), 1–31 (2024). https://doi.org/10.1145/3687060
Ennaji, M.: Multiculturalism, citizenship, and education in Morocco. In: Educational Scholarship across the Mediterranean, pp. 237–258. Brill (2021). https://doi.org/10.1163/9789004506602_017
Es-sousy, A., El Alaoui, Y., Jalid, K.: Exploring the challenges and strategies of interpreting humor in conferences "Arabic–English.". Int. J. Lang. Lit. Stud. **6**(4), 1–15 (2024) https://doi.org/10.36892/ijlls.v6i4.1968
Frey, C.B., Llanos-Paredes, P.: Lost in Translation: AI's Impact on Translators and Foreign Language Skills. VoxEU.org. Available https://cepr.org/voxeu/columns/lost-translation-ais-impact-translators-and-foreign-language-skills (2025). Last accessed 01 Jan 2026
Gajšt, N.: Applications of AI-driven tools in translating and drafting commercial correspondence—a Slovenian-English perspective. ELOPE: Engl. Lang. Overseas Perspect. Enquiries. (2025). https://doi.org/10.4312/elope.22.1.153-170
Gao, W.W., Kang, H.: Power distance orientation and LMX quality on workplace voice in China. J. Eurasian Stud. **16**(3), 141–168 (2019) https://doi.org/10.31203/aepa.2019.16.3.008
Geiger, I.: Communication media and negotiation: a review. In: Kilgour, D.M., Eden, C. (eds.) Handbook of Group Decision and Negotiation, pp. 1–20. Springer International Publishing, Cham (2019). https://doi.org/10.1007/978-3-030-12051-1_37-1
Genovese, A., Borna, S., Gomez-Cabello, C., Haider, S., Prabha, S., Forte, A., Veenstra, B.: Artificial intelligence in clinical settings: a systematic review of its role in language translation and interpretation. Ann. Transl. Med. **12** (2024) https://doi.org/10.21037/atm-24-162
Gonçalves, K., Schluter, A.: "Please do not leave any notes for the cleaning lady, as many do not speak English fluently": policy, power, and language brokering in a multilingual workplace. Lang. Policy. **16**(3), 241–265 (2017). https://doi.org/10.1007/s10993-016-9406-2
Guillouet, L., Khandelwal, A., Macchiavello, R., Teachout, M.: Language barriers in multinationals and knowledge transfers. SSRN Electron. J. (2021). https://doi.org/10.3386/W28807
Güven, Ç., Alishahi, A., Brighton, H., Nápoles, G., Olier, J.S., Šafář, M., Postma, E., Shterionov, D., De Sisto, M., Vanmassenhove, E.: AI in support of diversity and inclusion. arXiv preprint. (2025) https://doi.org/10.48550/arXiv.2501.09534
Harris-Watson, A.M., Larson, L.E., Lauharatanahirun, N., DeChurch, L.A., Contractor, N.: Social perception in human-AI teams: warmth and competence predict receptivity to AI teammates. Comput. Hum. Behav. **145**, 107765 (2023). https://doi.org/10.1016/j.chb.2023.107765
Harzing, A.-W., Feely, A.J.: The language barrier and its implications for HQ-subsidiary relationships. Cross cult. Manag.: Int. J. **15**(1), 49–61 (2008). https://doi.org/10.1108/13527600810848827
Harzing, A.-W., Köster, K., Magner, U.: Babel in business: the language barrier and its solutions in the HQ-subsidiary relationship. J. World Bus. **46**(3), 279–287 (2011). https://doi.org/10.1016/j.jwb.2010.07.005
Heidegger, M.: The Question Concerning Technology, and Other Essays. W. Harper & Row, New York, Translated by Lovitt (1977)
Hofstede, G.: Culture's Consequences: Comparing Values, Behaviors, Institutions, and Organizations across Nations, 2nd edn. Sage Publications, Thousand Oaks (2001)
Ishikawa, T.: English as a multilingua franca and 'trans-' theories. Englishes Pract. **5**(1), 1–27 (2022). https://doi.org/10.2478/eip-2022-0001
Ishkhanyan, M.: The sovereignty–internationalism paradox in AI governance: digital federalism and global algorithmic control. Discov. Artif. Intell. **5** (2025). https://doi.org/10.1007/s44163-025-00374-x
Jeong, H., Lindemann, S.: Facilitating or compromising inclusion? Language policies at Swedish higher education institutions as workplaces. Multilingua. **43**(3), 365–395 (2024). https://doi.org/10.1515/multi-2023-0077

Kalra, K., Danis, W.: Language and identity: the dynamics of linguistic clustering in multinational enterprises. J. World Bus. **59**(4), 101541 (2024). https://doi.org/10.1016/j.jwb.2024.101541

Kapoor, S.: The evolving role of human-in-the-loop evaluations in advanced AI systems. Eur. J. Comput. Sci. Inf. Technol. **13**(9), 115–126 (2025) https://doi.org/10.37745/ejcsit.2013/vol13n9115126

Karanges, E., Johnston, K., Beatson, A., Lings, I.: The influence of internal communication on employee engagement: a pilot study. Public Relat. Rev. **41**(1), 129–131 (2015). https://doi.org/10.1016/j.pubrev.2014.12.003

Kim, H., Yi, H.: Machine-facing English: defining a hybrid register shaped by human–AI discourse. arXiv preprint. (2025) https://doi.org/10.48550/arXiv.2505.2303

Koech, D.K., Degago, E., Kipkorir, C.S.S., Chittoo, H.B., Szabó, A.P., Molnár, E.: Internationalization and globalization in higher education: an insight on effect of machine translators on team performance among multicultural students working and studying in Hungary. J. Ecohumanism. **3**(4), 1–15 (2024) https://doi.org/10.62754/joe.v3i4.3575

Koskinen, K.: Translatoriality and technological mediation: machine translation as a cultural and ethical challenge. Transl. Stud. **13**(4), 446–461 (2020)

Kranzberg, M.: Technology and history: "Kranzberg's laws". Technol. Cult. **27**(3), 544–560 (1986). https://doi.org/10.2307/3105385

Kroskrity, P.V.: Language ideologies. In: Duranti, A. (ed.) A Companion to Linguistic Anthropology, pp. 496–517. Blackwell, Oxford (2004)

Kubota, R.: 'Language is only a tool': Japanese expatriates working in China and implications for language teaching. Multiling. Educ. **3**(4), 1–20 (2013). https://doi.org/10.1186/2191-5059-3-4

Leal, A.: The European Union's translation policies, practices and ideologies: time for a translation turn. Perspecta. **30**(2), 195–208 (2021). https://doi.org/10.1080/0907676X.2021.1927121

Lee, C.-S.: Exploring technological determinism in technology philosophy. J. Korean Pract. Arts Educ. **35**(3), 57–72 (2022) https://doi.org/10.24062/kpae.2022.35.3.57

Li, X.: Research on cultural subjectivity challenges and strategies for AI translation in the era of large language models. Engl. Lang. Teach. Linguist. Stud. **6**(6), 114–125 (2025) https://doi.org/10.22158/eltls.v6n6p114

Li, Z., Freeborn, R.: Reconfiguring digital accountability: AI-powered innovations and transnational governance in a postnational accounting context. arXiv preprint. https://arxiv.org/abs/2507.00288 (2025). https://doi.org/10.48550/arXiv.2507.00288

Liddicoat, A.: Intercultural mediation, intercultural communication and translation. Perspecta. **24**(3), 354–364 (2016). https://doi.org/10.1080/0907676X.2014.980279

Lindhout, P., Swuste, P.H.J.J., van Gulijk, C., Zwaard, W., Ale, B.J.M.: Learning from language problem related accident information in the process industry: a literature study. Process. Saf. Environ. Prot. **130**, 37–47 (2019). https://doi.org/10.1016/j.psep.2019.06.017

Long, H.: Learning to live with machine translation. New Lit. Hist. **54**(4), 721–753 (2022). https://doi.org/10.1353/nlh.2022.a898327

Lønsmann, D., Kraft, K.: Language policy and practice in multilingual production workplaces. Multilingua. **37**, 403–427 (2018). https://doi.org/10.1515/multi-2017-0088

López-Arroyo, B., Pizarro-Sánchez, I., Moreno-Pérez, L.: Sharing knowledge without borders: collaboration between U.S. business and law students and Spanish translation students. J. Virtual Exch. **7**, 60–79 (2024) https://doi.org/10.21827/jve.7.39420

Lucy, J.A.: Linguistic relativity. Annu. Rev. Anthropol. **26**, 291–312 (1997). https://doi.org/10.1146/annurev.anthro.26.1.291

Maitland, S., Tietze, S., Heath, D.M.: Multinational corporations as cultural translators: interpreting difference in diversity and inclusion. J. Transl. Lang. Stud. **3**(2), 1–18 (2022) https://doi.org/10.48185/jtls.v3i2.525

Massey, G.: Translation, diversification and corporate communications. Lang. Matters. Available https://blog.zhaw.ch/languagematters/2021/04/29/translation-diversification/ (2021). Last accessed 01 Jan 2026

Matsh: The Hidden Cost of "English-Only" Training in North African Factories. Available https://www.matsh.co/en/the-hidden-cost-of-english-only-training-in-north-african-factories/ (2025). Last accessed 01 Jan 2026

Mei, A.: How Does Language Influence our Minds? Public Media, From a Linguistics Perspective. Lect. Notes Educ. Psychol (2024) https://doi.org/10.54254/2753-7048/42/20240840

Merchant, A., Adnan, M.: The role of pragmatics in cross-cultural communication. Adv. Soc. Sci. Arch. J. **4**(1), 81–98 (2025) https://doi.org/10.55966/assaj.2025.4.1.081

Miller, C., Silverman, D., Jurica, V., Richerson, E., Morris, R., Mallard, E.: Embedding register-aware MT into the CAT workflow. In: Proc. 13th Conf. Association for Machine Translation in the Americas (Vol. 2: User Track), pp. 275–282. AMTA. Available https://aclanthology.org/W18-1920/ (2018). Last accessed 01 Jan 2026

Mohamed, Y.A., Kannan, A., Bashir, M., Mohamed, A.H.H.M., Adiel, M.A.E., Elsadig, M.A.: The impact of artificial intelligence on language translation: a review. IEEE Access. **12**, 25553–25579 (2024). https://doi.org/10.1109/ACCESS.2024.3366802

Moorkens, J.: Under pressure: translation in times of austerity. Perspect. Stud. Translatol. **25**(3), 464–477 (2017). https://doi.org/10.1080/0907676X.2017.1285331

Moorkens, J., Way, A., Lankford, S.: Sociotechnical effects of machine translation. arXiv preprint. (2025) https://doi.org/10.48550/arXiv.2503.20959

Muñoz-Basols, J., Neville, C., Lafford, B., Godev, C.: Potentialities of applied translation for language learning in the era of artificial intelligence. Hispania. **106**(1), 171–194 (2023). https://doi.org/10.1353/hpn.2023.a899427

Neeley, T.: The Language of Global Success: How a Common Tongue Transforms Multinational Organizations. Princeton University Press (2017)

Ningrum, D.K.S., Sofyan, R., Saragih, E.: Exploring linguistic competence in academic text translation by professionals. Lang. Literacy. **8**(1), 1–14 (2024) https://doi.org/10.30743/ll.v8i1.9277

Occupational Safety and Health Administration: OSHA Training Standards Policy Statement (Memorandum). Available https://www.osha.gov/laws-regs/standardinterpretations/2010-04-28 (2010). Last accessed 01 Jan 2001. (OSHA)

Occupational Safety and Health Administration: Training Requirements in OSHA Standards (OSHA 2254-09R 2015), U.S. Department of Labor. Available https://www.osha.gov/sites/default/files/publications/osha2254.pdf (2015). Last accessed 01 Jan 2026. (OSHA)

Occupational Safety and Health Administration: OSHA Resource for Development and Delivery of Training to Workers (OSHA 3824-05R 2021). U.S, Department of Labor (2021)

OECD: AI language models. OECD Digital Economy Papers. **366** (2023). https://doi.org/10.1787/13d38f92-en

Olohan, M.: Post-editing: a genealogical perspective on translation practice. In: Bisiada, M. (ed.) Empirical Studies in Translation and Discourse, pp. 1–25. Language Science Press, Berlin (2021). https://doi.org/10.5281/zenodo.4450077

Posey, C., Bennett, R.J., Roberts, T.L.: Understanding the mindset of employees in information security: a psychological perspective. Comput. Secur. **29**(4), 487–495 (2010). https://doi.org/10.1016/j.cose.2009.12.005

Rajagopal, S., Lakshmi, P.: Integrating AI in organizational management: implications for communication strategies and social dynamics. J. Inf. Syst. Eng. Manag. **10**(43s), 1–12 (2025) https://doi.org/10.52783/jisem.v10i43s.8408

Robertson, S., Díaz, M.: Understanding and being understood: user strategies for identifying and recovering from mistranslations in machine translation-mediated chat. In: Proc. ACM Conf. Fairness, Accountability, and Transparency, pp. 1–12 (2022). https://doi.org/10.1145/3531146.3534638

Roncevic, I.: Multilingual habitus and language prestige: value profiles of Croatian, English, German, and Italian in Croatia. World J. Engl. Lang. (2024). https://doi.org/10.5430/wjel.v14n5p98

Sadiqzade, Z.: The linguistic expression of emotion: a cross-cultural analysis. EuroGlobal J. Linguist. Lang. Educ. (2025) https://doi.org/10.69760/egjlle.2500195

Sandrini, P.: Beyond the cloud: assessing the benefits and drawbacks of local LLM deployment for translators. Open Transl. Stud. J. **8**(1), 1–15 (2025)

Sangwa, S., Murungu, E., Ssempa, M., Ekosse, E., Ruhamya, P.: Bridging cultures and codes: rethinking communication strategies in Africa-Europe business through AI and cultural intelligence. SSRN Electron. J. (2025). https://doi.org/10.2139/ssrn.5286085

Santorelli, M., Catullo, D.: Human mobility and language: towards new multilingual approaches with AI. Econ. Growth Cond. Globalization III. **1–12** (2023) https://doi.org/10.36004/nier.cdr.v.2023.17.16

Seyidov, R.: Importance and role of artificial intelligence for Arabic translators. Int. J. Arab. Linguist. Lit. Stud. **6**(2), 45–60 (2024) https://doi.org/10.31559/jalls2024.6.2.4

Shahmerdanova, R.: Artificial intelligence in translation: challenges and opportunities. Acta Globalis Humanitatis et Linguarum. **2**(1), 62–70 (2025) https://doi.org/10.69760/aghel.02500108

Śliwa, M., Johansson, M.: How non-native English-speaking staff are evaluated in linguistically diverse organizations: a sociolinguistic perspective. J. Int. Bus. Stud. **45**(9), 1133–1151 (2014). https://doi.org/10.1057/JIBS.2014.21

Szczepaniak-Kozak, A., Wąsikiewicz-Firlej, E., Lankiewicz, H.: Global English versus local language during a sojourn abroad: a narrative study. J. Intercult. Commun. **22**(2), 1–16 (2022) https://doi.org/10.36923/jicc.v22i2.58

Tarafdar, M., Cooper, C.L., Stich, J.F.: The technostress trifecta: techno eustress, techno distress, and design: theoretical directions and an agenda for research. Inf. Syst. J. **29**(1), 6–42 (2019). https://doi.org/10.1111/isj.12169

Tenzer, H., Schuster, T.: Language barriers in different forms of international assignments. In: Bader, B., Schuster, T., Bader, A.K. (eds.) Expatriate management: Transatlantic dialogues, pp. 63–100. Palgrave Macmillan (2017). https://doi.org/10.1057/978-1-137-57406-0_3

Tenzer, H., Pudelko, M., Harzing, A.-W.: The impact of language barriers on trust formation in multinational teams. J. Int. Bus. Stud. **45**(5), 508–535 (2014). https://doi.org/10.1057/jibs.2013.64

Tonkin, H.: Making sense of language in international affairs. In: Barnaby, F. (ed.) Building a more Democratic United Nations, pp. 207–210. Routledge, London (2020). https://doi.org/10.4324/9781003062493-47

Vanmassenhove, E.: On the integration of linguistic features into statistical and neural machine translation. arXiv preprint. (2020) https://doi.org/10.48550/arXiv.2003.14324

Vössing, M., Kühl, N., Lind, M., Satzger, G.: Designing transparency for effective human-AI collaboration. Inf. Syst. Front. **24**(4), 877–895 (2022). https://doi.org/10.1007/s10796-022-10284-3

Wilmot, N., Vigier, M., Humonen, K.: Language as a source of otherness. Int. J. Cross Cult. Manag. **23**(4), 1–18 (2023). https://doi.org/10.1177/14705958231216936

Winner, L.: Do artifacts have politics? Daedalus **109**(1), 121–136. Available https://www.jstor.org/stable/20024652 (1980). Last accessed 14 Jan 2026

Worakittikul, W., Mangdindam, T., Ketkaew, C., Naruetharadhol, P.: Breaking barriers: examining the impact of language on employee-driven innovation in European multinational subsidiaries in Thailand. Cogent Arts Humanit. **12**(1) (2025). https://doi.org/10.1080/23311983.2025.2451492

Xiao, A., Hancock, J.: Sustaining human agency, attending to its cost: an investigation into generative AI design for non-native speakers' language use. In: Proc. CHI Conf. Human Factors in Computing Systems, pp. 1–14 (2025). https://doi.org/10.1145/3706598.3713626

Yun, D., Jung, H.: Anger expression in negotiation: the effects of communication channels and anger intensity. Front. Psychol. **13** (2022). https://doi.org/10.3389/fpsyg.2022.879063

Zakhir, M., O'Brien, J.: Moroccan Arabic: the battlefield of language ideologies. Jurnal Arbitrer. **6**(1), 59–76 (2019) https://doi.org/10.25077/AR.6.1.59-76.2019

Zaki, M.Z., Ahmed, U.: Bridging linguistic divides: the impact of AI-powered translation systems on communication equity and inclusion. J. Transl. Lang. Stud. **5**(2), 25–42 (2024) https://doi.org/10.48185/jtls.v5i2.1065

Zavala-Rojas, D., et al.: Using machine translation and post-editing in the TRAPD approach for survey translation. UPF Repository. Available https://repositori.upf.edu/bitstreams/14b1dbfb-534e-4460-9c50-2d3e0c636cc4/download (2024). Last accessed 01 Jan 2026

Zhang, Y., Owusu, D.A., Carpuat, M., Gao, G.: Facilitating global team meetings between language-based subgroups: when and how can machine translation help? Proc. ACM Hum.-Comput. Interact. **6**(CSCW1), 1–26 (2022). https://doi.org/10.1145/3512937

The Use of Artificial Intelligence in Legal Translation: The Case of Sworn Translators in Morocco

Karima Bouziane and Abdelmounim Bouziane

Abstract This article investigates the impact of artificial intelligence (AI) on legal and official translation, with particular attention to the role of sworn translators in Morocco. It compares human and AI-assisted translations of Moroccan legal texts, evaluating them on efficiency, accuracy, and speed. The methodology involves a comparative analysis of outputs produced by AI tools and traditional translation practices, which rely on specialized legal dictionaries, glossaries, and professional expertise. Results reveal that AI significantly enhances translation speed and accessibility but struggles with domain-specific terminology and the subtle and context-specific expressions characteristic of Moroccan law. Subtle inaccuracies in translating legal concepts may compromise reliability and legal certainty if AI outputs are used independently. The findings emphasize the need for a hybrid translation approach, where AI serves as a supportive tool while human translators maintain oversight, validation, and certification. Such an approach ensures terminological precision, compliance with legal norms, and the overall quality and legitimacy of official translations.

Keywords Artificial intelligence · Legal translation · Sworn translators

1 Introduction

The integration of artificial intelligence (AI) into translation practices is transforming the field of legal and official translation worldwide. Legal translation, particularly in multilingual contexts like Morocco, demands not only linguistic

K. Bouziane
ALCS Lab, Faculty of Letters and Humanities, Chouaib Doukkali University, El Jadida, Morocco

A. Bouziane (✉)
Chouaib Doukkali University & Urban Agency of Casablanca, El Jadida, Morocco

A. Bouziane and K. Bouziane (eds.), *AI, Management, and Culture: Shaping Organizations in the Digital Age*, Studies in Computational Intelligence 1291,
https://doi.org/10.1007/978-3-032-27882-1_12

fluency but also precise adherence to specialized terminology, culturally embedded legal concepts, and formalized legal frameworks. Errors in this domain can compromise legal certainty, obstruct access to justice, and create professional and institutional risks. Recent advances in AI, including neural machine translation (NMT) and large language models (LLMs), offer opportunities to improve efficiency, accessibility, and consistency in translation tasks. However, studies have consistently shown that AI alone struggles with domain-specific terminology and nuanced legal contexts, highlighting the continued importance of human expertise.

In Morocco, the role of sworn translators is particularly critical due to the interplay of Arabic, French, and English in legal texts, as well as the influence of both Islamic Maliki jurisprudence and French legal traditions. While AI tools can expedite translation processes and assist translators in drafting, there is limited empirical evidence on how AI performs relative to human expertise in the Moroccan legal context. This study addresses this gap by comparing human and AI-assisted translations of Moroccan legal texts, evaluated by certified sworn translators for accuracy, legal fidelity, and usability. Through exploring the complementary roles of AI and human translators, this research provides insights into hybrid translation models that can enhance efficiency without compromising legal reliability.

2 Literature Review

The integration of artificial intelligence (AI) into legal translation has emerged as a significant development in translation studies, reflecting broader trends in digital transformation across legal systems worldwide. Scholars have increasingly examined the potential and limitations of AI-based translation tools, particularly neural machine translation (NMT) and large language models (LLMs), in handling legal texts where precision, terminological consistency, and contextual understanding are paramount (Khan 2025; Bajčić and Golenko 2024; Qi 2025). Legal translation, unlike general translation, involves domain-specific terminology, formalized language structures, and culturally embedded legal concepts. As such, errors in translation can have serious implications for legal certainty and access to justice (Briva-Iglesias 2021; Wright 2025).

Comparative studies examining human and AI translation of legal texts consistently highlight the advantages of human translators in managing terminological precision and contextual subtleties. Altakhaineh et al. (2025) demonstrated that human translations of Arabic legal documents outperformed AI systems such as ChatGPT in terms of accuracy, clarity, and adherence to formal legal frameworks. Similarly, Abdelaal and Al Sawi (2025) found that while NMT and LLM systems provided efficient and contextually relevant translations, human translators maintained superiority in fluency, legal terminology, and stylistic appropriateness. These findings suggest that AI tools, despite their computational efficiency, are insufficient when used independently in specialized legal domains where the stakes of misinterpretation are high.

Several studies have emphasized the potential of hybrid approaches, where AI acts as an assistive tool under the supervision of human translators. Alghamdi and Alotaibi (2025) explored the application of ChatGPT in translation quality assessment (TQA) and found that AI could effectively detect errors and provide corrective suggestions, aligning closely with human instructor evaluations. Moneus and Sahari (2024) further argued that while AI excels at repetitive and formulaic translation tasks, human oversight remains critical for interpreting legal subtleties. The notion of a collaborative framework is also evident in the work of Meng et al. (2025), who proposed a multi-agent system combining AI-driven terminology extraction with human-in-the-loop supervision to ensure precision in multilingual legal translation. These hybrid models underscore the complementary strengths of AI and human expertise, particularly in contexts involving complex legal systems or under-resourced languages.

The challenge of legal terminology remains a recurring theme in AI-assisted translation research. Studies in Arabic, French, and other languages reveal that AI often produces literal or culturally inappropriate translations of legal concepts, failing to capture local legal subtleties (Ait ElFqih and Monti 2023; Badah et al. 2026; Al-Maktary et al. 2025). For example, Bouziane and Bouziane (2025) noted difficulties in translating financial and audit-related legal terminology with ChatGPT, emphasizing that domain-specific knowledge and context awareness are crucial for accuracy. Similar observations were reported by Greńczuk et al. (2024) highlighting the limitations of DeepL, Google Translate, and other AI tools in producing precise translations for specialized legal texts. These findings confirm that AI, while capable of improving efficiency and accessibility, cannot fully replace human judgment in legal translation tasks, particularly for sworn translators who operate within formalized legal frameworks.

Further studies in the Moroccan context, the role of sworn translators is particularly significant due to the multilingual nature of legal drafting, often involving Arabic, French, and English (El Ghazi 2022). Recent studies have highlighted the growing integration of AI within Moroccan legal systems, including applications in criminal record processing (El Moussaoui et al. 2025) and legal question answering (Badah and Ghogho 2025). While AI can facilitate document processing and support decision-making, research indicates that Moroccan legal terminology, influenced by Islamic Maliki jurisprudence and French legal conventions, presents challenges for automated translation systems (Bahaj and Ghogho 2025). Thus, the effectiveness of AI-assisted translation tools is contingent upon domain-specific adaptation, oversight, and culturally informed post-editing.

Several scholars have also addressed broader implications of AI integration for legal translation, including ethical, institutional, and policy considerations. Khan (2025) highlighted concerns over accountability, employment, and the risks of bias in AI-assisted legal translation, advocating for people-involving workflows and robust quality assurance. Similarly, Clément-Wilz et al. (2025) emphasized that while AI enhances productivity and supports translation tasks in the European Union, human oversight remains indispensable for ensuring terminological consistency, cultural fidelity, and compliance with legal standards.

In summary, the literature demonstrates that AI-assisted translation holds considerable promise for improving efficiency, consistency, and access in legal translation. However, persistent challenges exist in managing domain-specific terminology, contextual specificities, and culturally embedded legal concepts, especially in multilingual and under-resourced contexts. Human translators, particularly sworn translators, remain essential to ensure translation accuracy, legal validity, and adherence to local legal practices. This duality has led scholars to advocate hybrid models, integrating AI as an assistive tool within human-supervised workflows, while emphasizing the importance of continuous evaluation and domain-specific training.

Despite the growing body of research on AI in legal translation, few studies have specifically investigated the interaction between AI tools and sworn translators in Morocco. While general comparative studies exist for Arabic legal translation (Altakhaineh et al. 2025; Abdelaal and Al Sawi 2025), there is a lack of empirical research assessing AI's performance relative to human expertise in the Moroccan legal system, which uniquely blends Arabic, French, and English legal traditions. Moreover, limited research explores the practical integration of AI into sworn translators' workflows, including efficiency, accuracy, and legal certainty in real-world translation tasks. This study aims to address these gaps by providing a context-specific analysis of AI-assisted translation in Morocco, emphasizing the need for hybrid approaches that preserve the essential role of human expertise.

3 Methodology

This study adopts a qualitative comparative research design to assess the performance of artificial intelligence in legal translation in comparison with human sworn translators in Morocco. The methodological approach is grounded in real translation practices and focuses on the criteria of accuracy, terminological precision, efficiency, and legal reliability, which are central to official and sworn translation.

3.1 Research Design and Corpus Selection

Two short but legally dense excerpts were selected from Moroccan legal and administrative texts commonly handled by sworn translators. The excerpts were chosen because they contain terminological, conceptual, and institutional references specific to the Moroccan legal system, making them particularly sensitive to translation inaccuracies.

The source texts were originally drafted in Arabic, reflecting the primary language of Moroccan legislation, and required translation into English, a frequent demand in international legal, commercial, and administrative contexts.

3.2 Translation Procedures

Two distinct translation processes were implemented:

Human Translation (HT)

The first translation was carried out by a certified sworn translator in Morocco. The translator relied exclusively on traditional professional resources, including:

- Official legal dictionaries (Arabic–French–English),
- Moroccan legislative texts and glossaries,
- Prior validated sworn translations,
- Personal professional expertise and contextual legal knowledge.
- No AI tools were used during this process.*AI-Assisted Translation (AIT)*

The same source text was translated using an AI-based large language model (ChatGPT) without post-editing or human intervention. The system was prompted to produce a formal legal translation suitable for official use.

3.3 Evaluation Process

The two translated versions (HT and AIT) were anonymized and submitted to two independent sworn translators, each with over ten years of professional experience. These evaluators were asked to assess the translations using a structured evaluation grid based on the following criteria:

- Terminological accuracy.
- Legal equivalence and conceptual fidelity.
- Linguistic clarity and formal register.
- Compliance with Moroccan legal usage.
- Overall suitability for official/legal use.

Evaluations were qualitative and comparative, focusing on identifying strengths, weaknesses, and risks associated with each translation.

3.4 Illustrative Example of the Translation Task

To illustrate the comparative translation process, two legally significant excerpts from Moroccan legislation were selected as representative examples. The first paragraph translated was as follows (Table 1):

The second legal paragraph translated was as follows (Table 2):

Table 1 Comparison of human and AI-assisted translation for contract approval text

Source text (Arabic) "يُسَلَّمُ هذا العقد بعد المصادقة عليه من طرف السلطات المختصة، ويُعتَبَرُ وثيقةً رسميةً تُثْبِتُ الالتزامات القانونية للأطراف وفقًا للتشريع المغربي الجاري به العمل."
Human translation (HT) "This contract, once approved by the competent authorities, shall be delivered as an official document establishing the legal obligations of the parties in accordance with the Moroccan legislation currently in force."
AI-assisted translation (AIT) "This contract is handed over after being validated by the relevant authorities and is considered an official document proving the legal commitments of the parties according to the applicable Moroccan law."

Table 2 Comparison of human and AI-assisted translation for personal data protection text

Source text 2 (Arabic) "تلتزم الأطراف بالامتثال لجميع أحكام التشريع المغربي فيما يتعلق بحماية البيانات الشخصية، ويجب على كل طرف اتخاذ التدابير اللازمة لضمان سرية المعلومات وعدم الإفصاح عنها لأي جهة ثالثة بدون موافقة صريحة."
Proposed human translation (HT) "The parties commit to comply with all provisions of Moroccan legislation concerning the protection of personal data, and each party must take the necessary measures to ensure the confidentiality of information and refrain from disclosing it to any third party without explicit consent."
Proposed AI-assisted translation (AIT) "The parties are obligated to follow all Moroccan laws regarding personal data protection, and each party should take measures to guarantee the confidentiality of information and not disclose it to any third party without clear consent."

4 Discussion

The results of this study confirm and extend previous findings on the use of AI in legal translation. Consistent with Altakhaineh et al. (2025) and Abdelaal and Al Sawi (2025), our study demonstrates that human translators outperform AI in maintaining terminological precision and legal fidelity, particularly in domain-specific contexts such as Moroccan law. The evaluations by sworn translators showed that AI-assisted translations, while fluent and readable, often produced subtle inaccuracies in legal terminology, such as "legal commitments" instead of "legal obligations," and "clear consent" instead of "explicit consent." These observations align with earlier studies (Bouziane and Bouziane 2025; Ait ElFqih and Monti 2023) highlighting AI's limitations in capturing culturally and legally embedded nuances.

At the same time, our results support the potential of AI to improve efficiency and accessibility, echoing findings by Alghamdi and Alotaibi (2025) and Moneus and Sahari (2024), who observed that AI can significantly reduce translation time and assist in repetitive or formulaic tasks. However, as emphasized by

Clément-Wilz et al. (2025) and Qi (2025), speed and fluency alone do not guarantee legal reliability. Our study reinforces the importance of human supervision, particularly in multilingual and hybrid legal systems such as Morocco's, which combine Arabic, French, and English legal terminology.

Overall, the findings underline the necessity of a hybrid translation model, integrating AI as a supportive tool under the oversight of experienced sworn translators. This approach not only leverages the efficiency gains offered by AI but also preserves accuracy, compliance, and legal certainty, addressing concerns raised in prior research on AI-assisted translation in specialized legal domains (Meng et al. 2025; Briva-Iglesias 2021). In the Moroccan context, where legal texts are highly specialized and multilingual, such a model is critical to maintaining the integrity of official translations while benefiting from technological advancements.

5 Results

The comparative evaluation revealed systematic differences between human and AI-assisted translations, particularly regarding legal precision and institutional conformity.

Accuracy and Terminological Precision

The human translation demonstrated high terminological consistency with Moroccan legal drafting conventions. Expressions such as "competent authorities" and "legislation currently in force" were evaluated as legally appropriate and aligned with sworn translation standards.

In contrast, the AI-assisted translation, while linguistically fluent, exhibited terminological approximations. Evaluators noted that the phrase "legal commitments" is less precise than "legal obligations", and that "applicable Moroccan law" lacks the normative specificity typically required in sworn translations. These subtle shifts were considered potentially problematic in official legal contexts.

Legal and Conceptual Fidelity

Both evaluators emphasized that human translation preserved the normative force and legal intent of the source text. The AI output, although semantically correct at a general level, was judged to reflect a functional equivalence rather than a strict legal equivalence. This distinction is critical in sworn translation, where wording may have direct legal consequences.

Linguistic Quality and Register

From a purely linguistic standpoint, the AI-assisted translation was considered clear, fluent, and stylistically appropriate. However, evaluators stressed that linguistic fluency alone is insufficient in legal translation when not supported by institutional and legal awareness.

Reliability for Official Use

Both sworn translators concluded that human translation was immediately usable for official and certified purposes, whereas the AI-assisted translation would require systematic human revision and validation before being legally acceptable. The absence of accountability and legal responsibility in AI output was highlighted as a major limitation.

Synthesis of Findings

Overall, the results confirm that AI tools offer significant gains in speed and accessibility, but they remain limited in handling Morocco-specific legal terminology and normative precision. The evaluators unanimously agreed that AI translations, if used without expert supervision, could introduce legal ambiguity and compromise legal certainty.

These findings strongly support a hybrid translation model, in which AI functions as a supportive tool for drafting or reference, while sworn translators retain full responsibility for validation, adaptation, and certification of legal texts.

6 Recommendations

The findings of this study suggest that artificial intelligence can serve as a valuable support tool in legal translation, but it cannot fully replace human expertise. Accordingly, institutions and translation professionals should adopt a hybrid approach, where AI assists in drafting, preliminary translation, or repetitive tasks, while sworn translators retain responsibility for reviewing, validating, and certifying the final output. This approach ensures that legal terminology, contextual nuances, and Moroccan legal conventions are accurately preserved, minimizing the risk of errors that could compromise legal certainty.

Training and capacity building are also critical to maximize the benefits of AI-assisted translation. Sworn translators should receive ongoing training in the use of AI tools, including how to integrate AI outputs with their professional judgment, verify terminology, and adapt translations to specific legal contexts. This dual competence, combining technological proficiency with legal expertise, can enhance efficiency without sacrificing quality.

At the institutional level, translation offices, law firms, and public authorities should establish quality assurance protocols that explicitly address the use of AI in legal translation. Such protocols could include standardized review procedures, terminological glossaries adapted to Moroccan law, and clear guidelines on accountability when AI tools are involved. By embedding these measures into professional workflows, organizations can ensure that AI contributes positively to translation practices while maintaining the legal reliability of sworn translations.

Finally, policymakers should consider regulatory and ethical frameworks that govern AI-assisted translation, particularly in multilingual and legally complex environments. Policies that promote transparency, accountability, and consistent terminology usage will help prevent misuse or overreliance on AI systems. Supporting research and collaboration between AI developers and legal translation experts can also foster the development of tools better tailored to the Moroccan legal system, ultimately enhancing both accessibility and quality in official translations.

7 Conclusion

The study demonstrates that AI-assisted translation can provide significant benefits in terms of speed, accessibility, and linguistic fluency, making it a useful tool for preliminary translation and drafting tasks. However, AI alone cannot fully capture the precise legal terminology, contextual nuances, and institutional conventions that characterize Moroccan legal texts. Evaluations by sworn translators revealed that while AI outputs are generally clear and readable, subtle inaccuracies in terminology and phrasing could compromise legal certainty if used without human oversight.

These findings highlight the necessity of a hybrid approach, where AI functions as an assistive tool, complementing but not replacing the expertise of sworn translators. Human oversight remains essential to ensure terminological accuracy, compliance with Moroccan legal norms, and the overall reliability of official translations. Training translators to work effectively with AI tools, implementing quality assurance protocols, and developing context-specific AI solutions are crucial steps to maximize the benefits of AI in legal translation. Ultimately, the integration of AI within a human-supervised framework offers a practical path to improving efficiency, supporting professional practice, and maintaining the legal integrity of translations in Morocco.

References

Abdelaal, N., Al Sawi, I.: A comparative evaluation of machine translation vs. human translation for legal texts: a case study of translations between English to Arabic. Comp. Legilinguistics. **63**, 186–223 (2025)

Ait ElFqih, K., Monti, J.: On the evaluation of terminology translation errors in NMT and PB-SMT in the legal domain: A study on the translation of Arabic legal documents into English and French. In: Proceedings of the Workshop on Computational Terminology in NLP and Translation Studies (ConTeNTS), pp. 26–35. INCOMA Ltd. (2023)
Alghamdi, F.A., Alotaibi, H.: Using AI in translation quality assessment: a case study of ChatGPT and legal translation texts. Electronics. **14**(19), 3893 (2025)
Al-Maktary, M.N., Mohammed, Z.A., Hassan, A.A., Almoughles, A.F., Almekdad, A.I.: Human versus artificial intelligence (AI): challenges in the legal translation of English to Arabic terminology at Yemeni universities. **8**(1), 177–212 (2025)
Altakhaineh, A.R.M., Alghathian, G.A., Jarrah, M.M.: A comparative study of accuracy in human vs. AI translation of legal documents into Arabic. Artif. Intell. Lang. Law. **14**, 120–142 (2025)
Badah, A.M., Khalaf, C.N., Dwaikat, F.J., AlQbailat, N.: The usage of artificial intelligence in legal translation: bridging the gap between law and language. Ampersand. **16**, 100248 (2026)
Bahaj, A., Ghogho, M.: MizanQA: benchmarking large language models on Moroccan legal question answering. arXiv preprint. https://arxiv.org/abs/2508.16357. (2025)
Bajčić, M., Golenko, D.: Applying large language models in legal translation: the state-of-the-art. J. Lang. Law. **13**, 171–196 (2024)
Bouziane, K., Bouziane, A.: AI-assisted translation of regulated financial documents: a case study on the auditor's report and memorandum of a French company. EDPACS. **1–9** (2025)
Briva-Iglesias, V.: Traducción humana vs. traducción automática: Análisis contrastivo e implicaciones para la aplicación de la traducción automática en traducción jurídica. Mutatis Mutandis. Revista Latinoamericana de Traducción. **14**(2), 571–600 (2021)
Clément-Wilz, L., de Laforcade, A., Cennamo, I.: Translation in the European Union: navigating technical and legal issues in the light of artificial intelligence—conclusions. Artif. Intell. Lang. Law. **14** (2025)
El Ghazi, O.: The legal translation profession in Morocco: perceptions of Moroccan sworn translators. Eur. J. Lang. Cult. Stud. **1**(2), 11–23 (2022)
El Moussaoui, T., Loqman, C., Boumhidi, J.: Decoding legal processes: AI-driven system to streamline processing of the criminal records in Moroccan courts. Intell. Syst. Appl. **25**, 200487 (2025)
Greńczuk, A., Chomiak-Orsa, I., Tryczyńska, K.: AI-supported translation tools for legal texts: a comparative analysis. Procedia Comput. Sci. **246**, 5545–5554 (2024)
Khan, M.I.: Impact of AI on legal and institutional translation: issues and strategies. Int. J. Transl. Interpret. Stud. **5**(6), 1–15 (2025)
Meng, L., Liu, M., Wang, H., et al.: Building from scratch: a multi-agent framework with human-in-the-loop for multilingual legal terminology mapping. Artif. Intell. Law. (2025)
Moneus, A.M., Sahari, Y.: Artificial intelligence and human translation: a contrastive study based on legal texts. Heliyon. **10**(6), e28106 (2024)
Qi, Y.: Challenges and opportunities of legal translation in the context of artificial intelligence. J. Innov. Dev. **11**(3), 220–233 (2025)
Wright, S.: Artificial intelligence in legal translation at the court of justice of the European Union. J. Lang. Law. **14**, 120–142 (2025)

Digital Transformation and Soft Skills Development in Moroccan Higher Education: Evidence from a Case Study at FSE-UM5

Mouna Darouassi and Oumelbanine El Gueddar

Abstract In the context of rapid digital transformation, higher education is increasingly expected to develop students' transversal skills (soft skills) alongside disciplinary knowledge. These skills, such as communication, collaboration, autonomy and critical thinking, have become essential for academic success and employability. This paper investigates how digital learning environments and educational technologies may support soft skills development among university students, while also considering challenges related to access and effective integration. The study is based on a survey conducted at the Faculty of Education Sciences of Rabat using a structured questionnaire addressing students' perceptions of digital tools and their role in fostering key soft skills. The paper discusses the main pedagogical and institutional implications of integrating digital technologies into higher education, and outlines recommendations for a more inclusive and effective adoption of educational digital practices.

Keywords Soft skills · Digital transformation · Higher education · Educational technology · Innovative pedagogy · Digital divide

M. Darouassi · O. El Gueddar (✉)
Politique Educatifs et Dynamiques Sociales (PEDS), Faculty of Education Sciences (FSE), Mohamed 5 University, Rabat, Morocco

M. Darouassi
Faculty of Education Sciences (FSE) PhD candidate and teaching monitor (monitrice) PEDS Laboratory, Mouna Darouassi Mohammed V University, Rabat, Morocco

O. El Gueddar
Faculty of Education Sciences (FSE) PEDS Laboratory, Oumelbanine El Gueddar, Rabat, Morocco

A. Bouziane and K. Bouziane (eds.), *AI, Management, and Culture: Shaping Organizations in the Digital Age*, Studies in Computational Intelligence 1291,
https://doi.org/10.1007/978-3-032-27882-1_13

1 Introduction

Higher education is currently experiencing a rapid transformation driven by the expansion of digital technologies, shifting labor market expectations, and the increasing demand for flexible learning modalities. In Morocco, this evolution aligns with national strategic priorities such as the New Development Model (2021) and the orientations of PACTESRI 2030, which emphasize innovation, inclusion, and quality improvement in university education. As universities seek to better prepare students for complex and dynamic professional environments, the focus has progressively shifted from knowledge transmission alone to the development of competencies that are transferable across contexts and careers.

Among these competencies, soft skills, including communication, collaboration, adaptability, critical thinking, and self-management are now widely recognized as essential for employability and long-term professional success. Recent global reports highlight that transversal competencies are increasingly valued across industries, particularly in contexts where automation and digitalization are reshaping job roles and required qualifications (OECD 2021; World Economic Forum 2023). In higher education, soft skills are also associated with academic success, as they contribute to student engagement, persistence, teamwork, and decision-making. However, integrating these skills into university curricula remains challenging, as they are often addressed implicitly rather than through structured and measurable learning strategies.

At the same time, digital learning environments offer new opportunities to foster soft skills development. Online platforms, interactive resources, and collaborative tools may support students' engagement, autonomy, and peer interaction, particularly when combined with active learning strategies. Research on educational technology suggests that digital tools can enhance different dimensions of student engagement, but their effects remain dependent on pedagogical design and contextual conditions (Bond et al. 2020). Moreover, digital learning effectiveness is influenced by the level of guidance provided to students and by the accessibility of digital resources, especially in contexts where disparities in connectivity and equipment persist (Hodges et al. 2020).

Despite the growing interest in digital education and soft skills, there remains limited empirical evidence documenting how university students perceive the role of digital tools in supporting soft skills development within the Moroccan higher education context. Addressing this gap is particularly important given the institutional shift toward digital transformation and competency-based education. Therefore, this study investigates students' perceptions of the contribution of digital technologies to the development of key soft skills, while also identifying the major barriers that may hinder effective and equitable digital integration in university training.

To achieve this objective, a quantitative survey was conducted among students at the Faculty of Education Sciences of Rabat (FSE-UM5). The questionnaire explored perceived digital impacts on engagement and motivation, communication and collaboration, autonomy and problem-solving, as well as access-related constraints. The findings of this study aim to provide pedagogical and institutional insights that may inform more inclusive and structured digital practices in higher education.

2 Literature Review

2.1 Soft Skills in Higher Education

Soft skills refer to a broad set of transversal competencies that shape how individuals interact, adapt, and perform in academic and professional contexts. These skills typically include communication, collaboration, critical thinking, creativity, self-management, and problem-solving. Unlike technical or discipline-specific skills, soft skills are transferable across tasks and environments, making them particularly relevant in higher education (Zarifian 2001). In fact, international reports increasingly emphasize the strategic role of soft skills in employability and workforce readiness, particularly in uncertain and fast-changing labor markets (OECD 2019; World Economic Forum 2023).

Despite their growing importance, integrating soft skills into university programs remains challenging. University curricula often prioritize disciplinary content and conventional assessment practices that do not easily capture interpersonal and behavioral competencies (European Commission 2016). As a result, soft skills are frequently developed implicitly through learning experiences rather than through structured educational strategies and explicit evaluation.

2.2 Digital Transformation and Learning Practices in Higher Education

Digital transformation has reshaped learning practices through the expansion of online platforms, learning management systems, interactive resources, and blended learning formats (Selwyn 2016). These developments have increased learning flexibility and enabled access to diverse content beyond the traditional classroom, redefining students' relationships with knowledge and instructional support.

However, research suggests that digital learning effectiveness depends heavily on how technologies are pedagogically integrated. Hodges et al. (2020) distinguish emergency remote teaching from well-planned online learning, highlighting that instructional design quality and learner support play a decisive role in educational outcomes. Furthermore, digital transformation should be understood as a pedagogical shift rather than a purely technical adoption process, as learning gains depend on alignment between digital tools and educational goals (Means et al. 2014).

2.3 Educational Technology and Student Engagement

Student engagement is widely recognized as a key predictor of academic achievement in higher education. Engagement generally includes behavioral participation, cognitive investment, and emotional involvement in learning tasks (Bond et al. 2020). Digital environments can support engagement by enabling flexible access to learning resources, increasing interactivity, and providing feedback opportunities that enhance student involvement.

In a systematic evidence map, Bond et al. (2020) show that educational technologies can influence multiple dimensions of student engagement; however, outcomes vary depending on learners' digital readiness, learning design, and the instructional context. Therefore, while digital tools may strengthen engagement, they are most effective when embedded in active learning approaches and structured instructional practices (Means et al. 2014).

2.4 Collaboration and Communication through Digital Tools

Collaboration and communication are among the most frequently emphasized soft skills in university learning environments. Digital tools can support teamwork by enabling peer-to-peer interaction through shared workspaces, discussion platforms, and online communication channels (Selwyn 2016). Such environments can facilitate coordination and knowledge sharing, particularly in project-based and collaborative learning activities.

Nevertheless, collaboration effectiveness depends largely on pedagogical structuring. Means et al. (2014) argue that online collaborative learning requires clear task organization, defined roles, and adequate guidance to produce meaningful interaction and learning outcomes. Without scaffolding, digital collaboration may lead to uneven participation, limited knowledge construction, or superficial exchanges.

2.5 Autonomy, Self-Regulation, and Problem-Solving in Digital Learning

Autonomy and problem-solving are key competencies in higher education, closely related to students' ability to self-manage learning processes and overcome academic difficulties. Digital learning environments can encourage autonomy by allowing students to access resources independently, revisit materials, and explore complementary content based on individual needs (Means et al. 2014).

However, autonomy requires more than access to information. It depends on self-regulation, time management, and critical evaluation of digital content. In this respect, digital learning may also create challenges such as information overload and distraction, which can weaken students' learning organization if guidance is insufficient (Selwyn 2016). Thus, digital tools can contribute to autonomy and problem-solving, but this contribution remains conditional upon students' digital literacy and the presence of instructional support.

2.6 Barriers to Digital Integration and the Digital Divide

Despite the opportunities created by educational technologies, barriers related to access remain significant. The digital divide may involve limited access to devices, unstable connectivity, restricted platform availability, or insufficient digital skills (OECD 2019). These constraints can reduce participation in digital learning activities, weaken learning continuity, and reinforce inequalities among students.

From an institutional perspective, equity in access is essential for inclusive digital integration. OECD (2019) emphasizes that digital learning reforms must consider both infrastructure and skills development to ensure that digital transformation benefits all learners rather than amplifying existing disparities.

2.7 Research Gap and Positioning of the Present Study

Overall, the literature suggests that digital learning environments can support soft skills development in higher education by fostering engagement, facilitating collaboration, and promoting autonomy. Yet, these benefits depend strongly on pedagogical design and on the availability of equitable learning conditions (Hodges et al. 2020; Means et al. 2014).

Within the Moroccan higher education context, further empirical research is needed to better document students' perceptions regarding the extent to which digital tools support soft skills development and the obstacles that may hinder inclusive integration. Therefore, the present study addresses this gap by investigating students' perceptions at FSE-UM5 through a quantitative survey focusing on engagement, communication and collaboration, autonomy and problem-solving, and access-related constraints.

3 Methodology

3.1 Research Design

This study adopted a quantitative research design to examine students' perceptions regarding the role of digital technologies in supporting soft skills development in higher education. A survey approach was selected as it enables the collection of standardized data and facilitates the identification of general trends across the participant group.

3.2 Research Setting and Participants

The study was conducted at the Faculty of Education Sciences of Rabat (FSE-UM5). Participants were recruited using a convenience sampling approach among students enrolled in different academic programs and levels. A total of 81 students voluntarily completed the questionnaire, representing diverse academic profiles and study tracks within the institution.

3.3 Instrument

Data were collected using a structured questionnaire designed to assess students' perceptions of the relationship between digital learning tools and the development of transversal competencies (soft skills). The instrument consisted mainly of closed-ended items addressing both the perceived contribution of digital technologies to key learning dimensions and the barriers that may limit effective digital integration.

3.4 Variables and Measures

The questionnaire examined four main dimensions:

- Engagement and motivation, referring to students' perceived involvement and willingness to participate in learning activities when using digital tools.
- Communication and collaboration, referring to students' perceptions of peer interaction and teamwork support through digital environments.
- Autonomy and problem-solving, referring to students' perceived ability to organize learning tasks, search for information, and manage academic challenges using digital resources.

- Access-related barriers, referring to difficulties in using digital tools due to connectivity, equipment, or resource availability constraints.

Different measurement scales were used depending on item characteristics. Demographic variables (e.g., gender, age) were measured using nominal categories. Students' reported frequency of participation and perceived relevance of selected competencies were assessed using ordinal response scales, ranging from "rarely" to "always". Perceived importance and perceived impact were measured using Likert-type scales, with response options ranging from "not important at all" to "very important".

3.5 Data Collection Procedure and Data Analysis

The questionnaire was administered to students at FSE-UM5. Participation was voluntary, and responses were collected for academic research purposes in order to capture students' perceptions of digital learning experiences within the university context. Data were coded and analyzed using descriptive statistics, primarily through frequencies and percentages (%), to summarize response patterns across the study dimensions. Results were reported using tables and graphical visualizations to facilitate interpretation and support the discussion of key findings.

3.6 Ethical Considerations

Participation in the study was voluntary and responses were collected anonymously. Participants were informed about the purpose of the research and provided informed consent prior to completing the questionnaire. The collected data were used exclusively for academic research purposes.

4 Results

4.1 Demographic Information of Respondents

See Fig. 1.

Table summary (Table 1):

Interpretation:

The sample is predominantly composed of female respondents (59.26%), compared to male respondents (40.74%), indicating a higher female participation rate in the survey (Fig. 2).

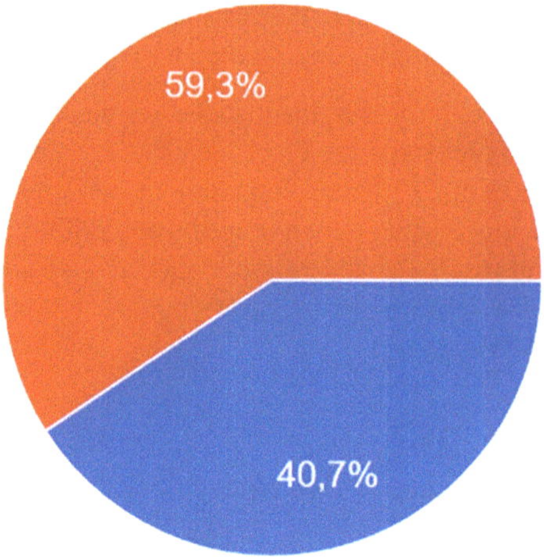

Fig. 1 Distribution of respondents by gender (N = 81)

Table 1 Distribution of respondents by gender (N = 81)

Gender	Percentage (%)
Female	59.26
Male	40.74
Total	**100.00**

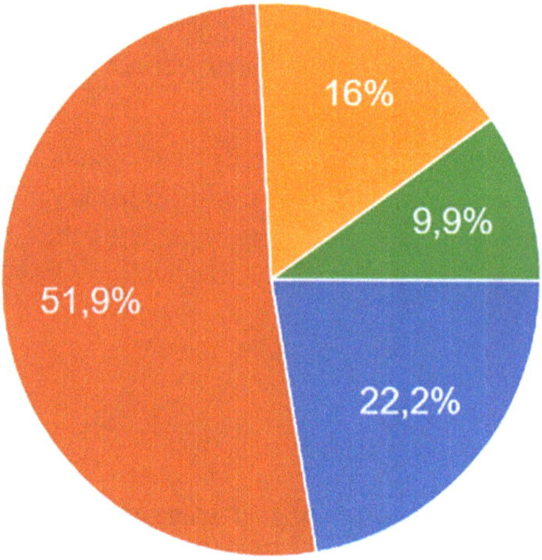

Fig. 2 Distribution of respondents by age group

Table summary (Table 2):

Interpretation:

The results show that most respondents are aged 22–25 years (51.85%), followed by the 18–21 age group (22.22%). The 26–30 and 31 and above categories are less represented, accounting for 16.05% and 9.88%, respectively. Overall, this distribution

Table 2 Distribution of respondents by age group (N = 81)

Age group	Percentage (%)
18–21	22.22
22–25	51.85
26–30	16.05
31 and above	9.88
Total	**100.00**

suggests that the survey findings mainly reflect the perceptions of young adults, particularly those aged 22–25 (Fig. 3).

Table summary (Table 3):

Interpretation:

The majority of respondents are enrolled in Master's programs (32.10%), followed by Excellence Bachelor's (22.22%) and Specialized Master's programs (17.28%). Professional Bachelor's and Education Bachelor's students represent 13.58% and

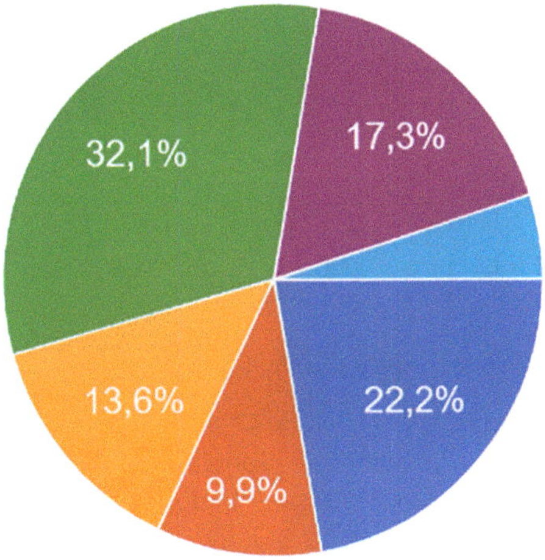

Fig. 3 Distribution of respondents by academic level

Table 3 Distribution of respondents by level of study (N = 81)

Level of study	Percentage (%)
Excellence Bachelor's	22.22
Education Bachelor's	9.88
Professional Bachelor's	13.58
Master's	32.10
Specialized Master's	17.28
PhD	4.94
Total	**100.00**

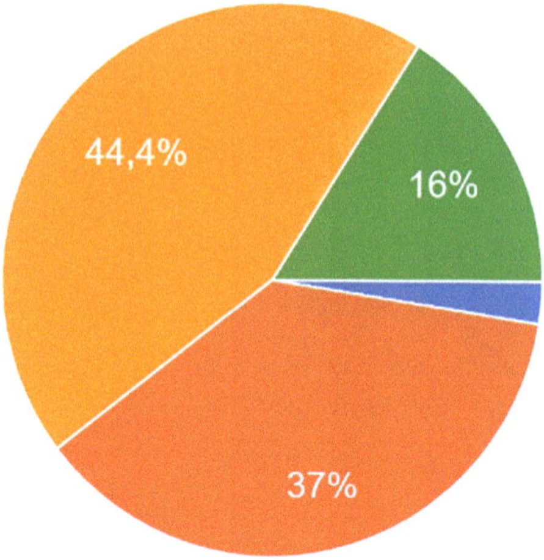

Fig. 4 Distribution of respondents by duration of study at FSE-UM5

Table 4 Distribution of respondents by years of study at FSE-UM5 (N = 81)

Years of study	Percentage (%)
Less than 1 year	2.47
1–2 years	37.04
2–4 years	44.44
More than 4 years	16.05
Total	**100.00**

9.88% of the sample, respectively, while PhD students are the least represented group (4.94%) (Fig. 4).

Table summary (Table 4):

Interpretation:

Most respondents report studying at FSE-UM5 for 2–4 years (44.44%), followed by 1–2 years (37.04%). Students with more than 4 years of study represent 16.05%, whereas those with less than one year at the institution are minimally represented (2.47%). This distribution indicates that the majority of participants have sufficient familiarity with the university learning environment to provide informed perceptions regarding digital learning practices.

4.2 *Effect of Digital Tools on Student Engagement and Motivation*

The findings indicate that 88% of the respondents perceive digital technologies as having a positive effect on their engagement and motivation during the learning

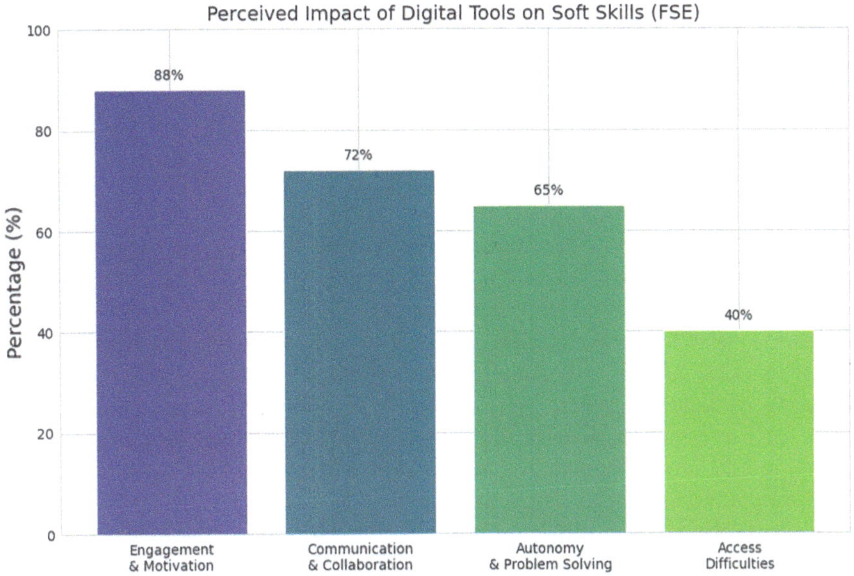

Fig. 5 Effect of digital tools on engagement and motivation

process. This suggests that digital learning environments are generally associated with increased student involvement, particularly through easier access to academic resources, flexible learning opportunities, and interactive digital content.

From an educational perspective, engagement is a fundamental factor for learning continuity and academic persistence. In this sense, students' perceptions reflect the potential of digital tools to foster a more active learning experience by enabling them to manage their academic activities more efficiently and remain connected to course materials beyond classroom time (Fig. 5).

4.3 Communication and Collaboration Through Digital Tools

Regarding social and collaborative competencies, 72% of the respondents report that digital tools improve communication and collaboration among students. This finding highlights the role of digital platforms and communication technologies in facilitating peer interaction, group coordination, and teamwork activities, especially in contexts where learning extends beyond face-to-face classroom sessions.

However, it is important to note that collaboration outcomes are not solely determined by the availability of technology. The effectiveness of digital collaboration is closely linked to how group activities are organized, the clarity of instructions, and the level of guidance provided within academic tasks. Therefore, digital tools may enable collaborative learning, but structured pedagogical practices remain essential to ensure meaningful interaction and teamwork skill development.

4.4 Digital Learning, Autonomy, and Problem-Solving

The results also show that 65% of students perceive digital learning as supporting their autonomy and problem-solving abilities. This may be explained by the fact that digital environments encourage self-directed practices such as searching for information, accessing diverse learning materials, and independently managing academic tasks.

Nevertheless, autonomy remains a complex competency that depends on students' ability to regulate their learning, organize time effectively, and critically evaluate digital information. As a result, the perceived contribution of digital tools to autonomy may vary depending on students' digital literacy and the extent of instructional support provided. This suggests that digital learning can promote autonomy most effectively when accompanied by guidance strategies that help learners build structured and sustainable independent learning habits.

4.5 Access Barriers and the Digital Divide

Despite the generally positive perceptions reported across soft skills dimensions, 40% of respondents state that they experience difficulties in accessing digital tools. This result points to the persistence of access-related barriers, such as insufficient equipment, unstable internet connectivity, or limited availability of digital resources.

These constraints may reduce students' participation in digital learning activities and create unequal learning conditions. Consequently, the findings underline that digital transformation in higher education is not only a pedagogical challenge but also an equity issue. Ensuring inclusive access to digital tools remains a key condition for maximizing the educational benefits of digital learning and supporting soft skills development across the student population (Fig. 6).

To provide a comprehensive overview of the findings, a radar profile was developed to compare the main dimensions simultaneously. This visualization summarizes the overall perceived impact of digital learning on soft skills, while also highlighting access-related constraints (Fig. 7).

Taken together, the results suggest that digital learning environments are perceived as supportive of several transversal competencies, while also revealing persistent access constraints that may limit students' ability to benefit equally from digital integration.

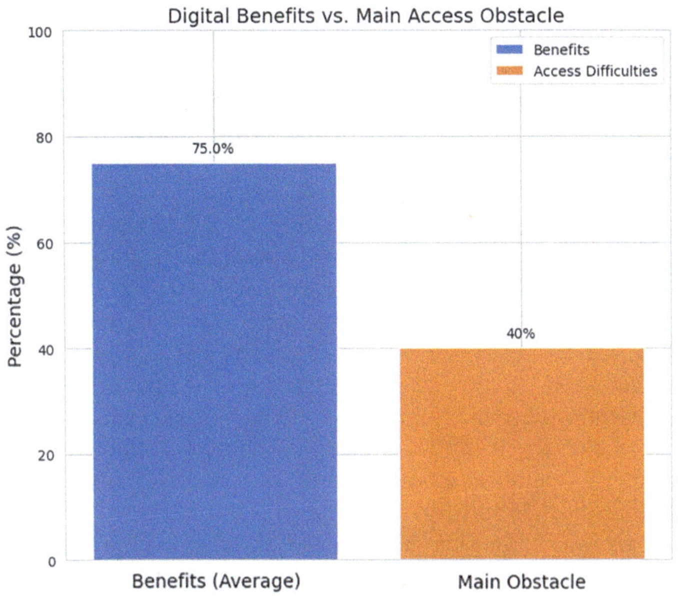

Fig. 6 Benefits versus access barriers to digital tools

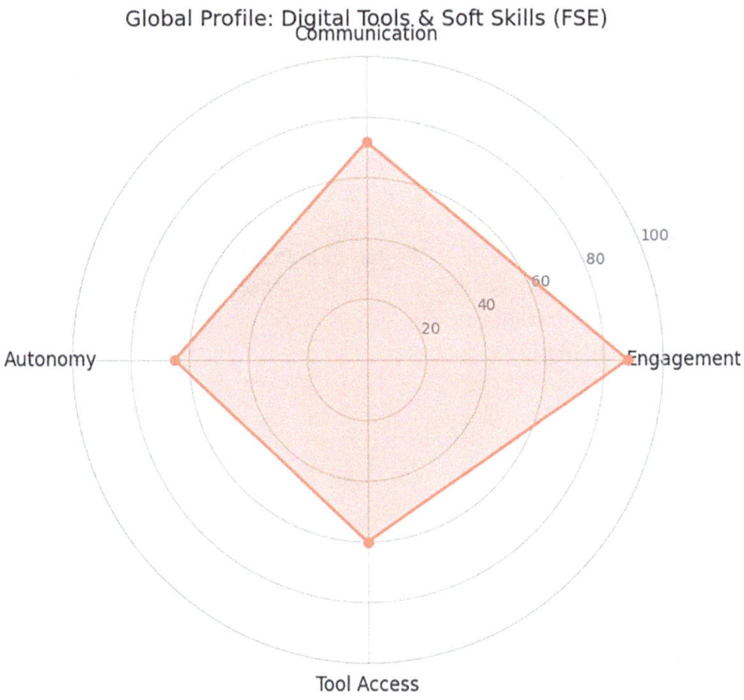

Fig. 7 Overall profile-digital learning and soft skills (radar chart)

5 Discussion

The findings highlight that digital technology has become a meaningful component of students' learning experiences in higher education. However, its perceived contribution to soft skills development should not be interpreted as a direct or automatic effect. Instead, the results suggest a differentiated pattern: digital tools appear to strongly support engagement and peer interaction, while their influence on autonomy and problem-solving remains more conditional and dependent on pedagogical and contextual factors. This distinction is important because it shifts the discussion from a technology-centered perspective to a learning-centered one, where outcomes depend on how digital environments are designed, supported, and accessed.

First, the strong association between digital learning and engagement may be explained by the flexibility and convenience offered by digital environments. Digital tools reduce organizational barriers by allowing students to access learning materials at any time, manage tasks more efficiently, and continue learning beyond classroom boundaries. Such features may increase students' perceived control over their learning process, which can enhance motivation and sustained participation. Nevertheless, engagement should not automatically be interpreted as deep learning. Digital engagement can reflect an immediate positive experience linked to accessibility and speed without necessarily ensuring long-term knowledge consolidation or higher-order skill development. This raises an essential pedagogical question: how can universities transform digitally supported engagement into meaningful cognitive involvement rather than short-term satisfaction?

Second, the perceived improvement in communication and collaboration confirms that digital tools can facilitate interaction among students by enabling constant connectivity through platforms, group spaces, and communication channels. Yet, this result also requires careful interpretation. Collaboration is not produced by technology alone; rather, it emerges from the way learning activities are structured. Digital tools make interaction possible, but effective teamwork requires clear task organization, shared objectives, and instructional guidance. Without such pedagogical framing, collaboration may become superficial or uneven, with limited contribution from some students, miscommunication, or fragmented group coordination. Therefore, digital environments can create collaborative opportunities, but the development of collaboration as a true soft skill depends on purposeful learning design and the quality of peer engagement.

Third, the more moderate effect observed for autonomy and problem-solving may reflect the complexity of these competencies. Autonomy requires students to plan learning activities, manage time, select relevant information, and persist through academic difficulties. Digital environments may support these practices by providing access to diverse resources and encouraging independent exploration. However, they may also create challenges such as information overload, distraction, or reliance on quick solutions. In this sense, digital learning can simultaneously enable and weaken autonomy depending on students' self-regulation capacities and the level of academic support provided. The findings therefore suggest that autonomy should be considered a competence that must be gradually built through guidance, scaffolding strategies, and information literacy training, rather than assumed as a natural outcome of digital access.

One of the most significant insights of this study is the persistence of access-related difficulties, which reveals that digital transformation remains constrained by material and infrastructural conditions. These barriers remind us that digital learning effectiveness is inseparable from equity considerations. When access to devices, connectivity, or digital resources is unequal, digital integration may reduce participation and reinforce existing disparities in learning opportunities. Consequently, the digital divide should not be treated as a purely technical issue; it becomes a pedagogical and social factor that directly shapes students' ability to benefit from digital learning and develop transversal competencies.

Overall, this study suggests that the value of digital learning for soft skills development depends on a combination of conditions, including accessibility, digital competence, pedagogical structuring, and institutional support. Digital tools represent a potential learning environment for competency development, but they can only produce consistent benefits when embedded within coherent educational strategies and inclusive policies. Future research could deepen this analysis by exploring differences across student profiles and by combining quantitative findings with qualitative data in order to better understand the mechanisms through which digital learning environments shape soft skills development in practice.

6 Conclusion

This study contributes to ongoing debates on competency-based higher education by highlighting the strategic relevance of digital learning environments in supporting transversal skill-building within university settings. Beyond technological adoption, the findings emphasize the need to approach digital transformation as an educational project that requires coherent instructional planning, learner support, and institutional responsibility. In this sense, digitalization should be viewed as a space for pedagogical redesign, where learning experiences are intentionally structured to cultivate communication, collaboration, initiative, and self-regulated learning.

From an institutional perspective, the study underlines that improving learning quality through digital practices cannot be separated from equity-oriented considerations. Strengthening the digital ecosystem of universities therefore involves not only expanding platforms and resources, but also ensuring that students can participate under fair and supportive conditions. Investing in infrastructure, digital literacy, and guided learning practices becomes essential to sustain educational effectiveness and reduce vulnerability to access constraints.

Finally, this research opens opportunities for further investigation through broader sampling across disciplines and institutions, as well as mixed-method approaches that combine student perceptions with qualitative insights and competency-based measurement. Such directions would allow deeper understanding of how digital practices shape students' learning behaviors over time and how higher education can operationalize soft skills development in a consistent, inclusive, and measurable manner.

Acknowledgments The author would like to express sincere gratitude to her thesis supervisor for guidance and support. The author is a PhD candidate and teaching assistant, supported by a monthly scholarship from the Moroccan National Center for Scientific and Technical Research.

Conflicts of Interest The author declares no conflict of interest.

Authenticity Statement This manuscript is an original work and has not been previously published or submitted for publication elsewhere.

Bibliography

Aldowah, H., Rehman, S.U., Ghazal, S., Umar, I.N.: Digital transformation in higher education: a systematic review. Educ. Inf. Technol. **29**(1), 1–28 (2024)

Billett, S.: Vocational Education: Purposes, Traditions and Prospects. Springer (2011)

Bond, M., Buntins, K., Bedenlier, S., Zawacki-Richter, O., Kerres, M.: Mapping research in student engagement and educational technology in higher education: a systematic evidence map. Int. J. Educ. Technol. High. Educ. **17**(2), 1–30 (2020)

Dede, C.: Comparing frameworks for 21st century skills. In: Bellanca, J., Brandt, R. (eds.) 21st Century Skills: Rethinking how Students Learn, pp. 51–76. Solution Tree Press (2010)

European Commission: A New Skills Agenda for Europe: Working Together to Strengthen Human Capital, Employability and Competitiveness. Publications Office of the European Union (2016)

Goleman, D.: Emotional Intelligence: why it Can Matter More than IQ. Bantam Books (1995)

Hodges, C., Moore, S., Lockee, B., Trust, T., Bond, A.: The difference between emergency remote teaching and online learning. EDUCAUSE. Review. (2020)

Jenkins, H.: Convergence Culture: where Old and New Media Collide. New York University Press (2006)

Kaplan, A.: Higher Education and the Future of Work: the Role of Soft Skills Development. Routledge (2021)

Kivunja, C.: Do you want your students to be job-ready with 21st century skills? Int. J. High. Educ. **3**(4), 81–91 (2014)

Means, B., Bakia, M., Murphy, R.: Learning Online: What Research Tells us about whether, when and how. Routledge (2014)

Nhedzi, A.: The challenges of online learning during COVID-19 pandemic in developing countries: students' perspective. Int. J. Educ. Res. Open. **2**, 100110 (2021)

OECD: OECD Skills Outlook 2019: Thriving in a Digital World. OECD Publishing (2019)

OECD: 21st-Century Readers: Developing Literacy Skills in a Digital World. OECD Publishing (2021)

Redecker, C.: European Framework for the Digital Competence of Educators (DigCompEdu). Publications Office of the European Union (2017)

Said-Hung, E., Marcano, B., Garay, U.: Digital divide and digital readiness in online education: a multidimensional approach. Comput. Educ.: Artif. Intell. **3**, 100072 (2022)

Selwyn, N.: Education and Technology: Key Issues and Debates, 2nd edn. Bloomsbury (2016)

Van Dijck, J., Poell, T., De Waal, M.: The Platform Society: Public Values in a Connective World. Oxford University Press (2018)

World Economic Forum: The Future of Jobs Report 2023. World Economic Forum (2023)
Zarifian, P.: Le modèle de la compétence: Trajectoire Historique, Enjeux Actuels et Propositions. Éditions Liaisons (2001)